Vibration Spectrum Analysis
A PRACTICAL APPROACH
Second Edition

Vibration Spectrum Analysis A PRACTICAL APPROACH

Second Edition

Steve Goldman
Goldman Machinery Dynamics Corporation
6 Mallard Drive
West Nyack, NY 10994
or
100 Gray Mouse Road
Saugerteis, NY 12477
Tele: 914-634-0674
Fax: 914-638-3859
Website: www.vibemaven.com
Email: sgoldman@vibemaven.com

Industrial Press Inc.

Library of Congress Cataloging-in-Publication Data

Goldman, Steve.
 Vibration spectrum analysis : a practical approach / Steve Goldman. — 2nd Ed.
  352 p. 15.6 x 23.5 cm.
  Includes index.
  ISBN 0-8311-3088-1 (alk. paper)
  1. Machinery–Vibration–Measurement.   1. Title.
 TJ177.G59   1999
 621.8′11–dc21   99-17685
                                                                                    CIP

Industrial Press Inc.
989 Avenue of the Americas
New York, N.Y. 10018

Second Edition

Copyright © 1999 by Industrial Press Inc., New York, New York.
Printed in the United States of America. All rights reserved. This book, or parts thereof, may not be reproduced, stored in a retrieval system, or transmitted in any form without the permission of the publishers.

10   9   8   7   6   5

Preface

This book has been written with several goals in mind. Stating what these goals are at the outset, as well as stating what the goals are not will allow the reader to decide how to make the best use of this book in one of several ways:

- Treat the book as a reference tool kept near instrumentation, ready to be brought into the dirt and liquid spray that makes up the environment of the machinery expert's battlefield.

- Use the book as a practical supplement to the many books that teach everything about the mathematics of vibration analysis and nothing about the practical matter of actually solving a problem, which may well be costing one's employer hundreds of thousands of dollars a day. This is the stuff of which heroes are made.

- Decide that the book is not sufficiently cluttered with formulas to have any use, and store it on a bookshelf where its lovely green binding may be admired by all who enter the office.

Vibration Spectrum Analysis grew out of a reluctance to travel. While I was working at Nash Engineering in the early 1970s, it became obvious that, unless I could train Nash's service technicians to do vibration analysis, I would have to travel throughout the known universe solving problems. The first thing I learned in this endeavor was that no textbook on vibration existed that did not assume a thorough knowledge of differential equations. I had to write my own training materials and train the Nash technicians myself. Soon, the only vibration problems that I had to deal with personally were the difficult ones, the politically "hot" ones, and the ones near New Orleans.

These early training efforts were extended by Nicolet Scientific, where it was obvious that, if only PhD's knew anything about vibration analysis, Nicolet would not sell many spectrum analyzers. The 3- to 5-day seminars I now give both to the public and in-house are a refinement of those early seminars for Nicolet, which tried to teach American Industry that there was a lot of potential money to be saved by doing sophisticated predictive maintenance and problem solving.

Of even more importance, it was shown that any competent mechanic, given the interest, could learn to become a "company hero" in the area of machinery reliability. With the reduction in downtime and repair costs, American Industry can again compete and, in fact, regain leadership in the production of heavy machinery and in processing items such as paper and chemicals. The high American wage can, at least partially, be offset by improved efficiency.

This book was written for the three types of people most likely to attend one of my seminars:

1. High school graduates, with little or no college, who have a good working knowledge of machinery in general.
   These are the people who keep American industry going.

2. Older college graduates, with many years of experience around heavy machinery and process plants. They have long since forgotten differential equations. Usually, these are the men who give instructions to the technicians mentioned above.

3. Recent college graduates, who lack many years of experience, but realize that their academic achievements are not sufficient to do well in the real world.

The book, then, was written to be light on mathematics, using it only where it facilitates a better understanding of the topic. An attempt was made to go heavy on the practical matters of developing an intuitive understanding of the nature of vibration, the way in which instrumentation can be used to deal with a problem, and, most importantly, the way in which one can proceed in attacking *any* problem with a good chance of success.

The book was created by rewriting the notes used in my seminars and trying to recall every question of interest ever asked in any of the seminars. Also, new sections were added to reflect improvements in the technology and instrumentation. In the belief that this author will not do a particularly good job of covering a tropic that does not interest him, only a cursory mention of modal analysis and a short appendix on balancing has been included. There are few case histories, because the author believes that the reader should use the information in this book to make his own case histories. Cookbook prescriptions make better cookies than they do competent analysis.

The information presented in this book is unlikely to become dated, with the possible exception of those few sections on available instrumentation. Note, however, that if one were to read a description of a Model A Ford, one would be able to figure out the important components of a new racing car. The appendices at the end of the book are based on previously published papers or talks to professional groups. Pick and choose among them as needed.

Lastly, this book is not intended to be used as an engineering college text. No attempt has been made to derive, for example, the Laws of Motion, because this would have been a distraction to the people for whom the book has been written. The technical school student may well find this book to be a chance at a well-paying, satisfying job upon graduation and the engineering college student may find it a good supplement to his theoretical studies in classical vibration analysis. Both types of student will have to overcome their revulsion at not having a great many homework problems at the back of each chapter.

If the reader has properly thought about this preface and decided to go on with the study of all or part of this book, the author bids them proceed in good health. If not, enjoy the green binding.

Steve Goldman, P.E.

Acknowledgements

This book is about vibration spectrum analysis in particular and about problem solving in general. My knowledge of vibration analysis is based on the patient instruction of Austin H. Church and Mike Rabins, both formerly of the now defunct New York University School of Engineering and Science. I had the privilege of learning from both these men as an undergraduate and, later, as a graduate student. I also wish to thank my parents, without whose support and encouragement my education would have been impossible.

Problem solving is more difficult to master than vibration analysis. It takes a lifetime of practice to develop. The payback is that it works in almost all areas of pursuit. For this ability, I first thank my father, Joe Goldman, for helping me to develop it over a period of many years. I was never given a solution, only help in addressing a problem. For help in slanting my problem solving ability toward commercial vibration problem solving, I am grateful for the friendship of the late Irving H. Owen, who allowed me to work on any project I chose at Nash, while being available for debate as to possible courses of action on a given problem.

Vibration analysis is not the most lucrative of pursuits, but it is considerably more fun than many forms of employment. I am, therefore, grateful to my wife Rosemary and to my children, Danny and Lizabeth for their support. Without this kind of cooperation, it would not be possible to continue to practice vibration analysis, the playground of my choice, in a spirit of family cooperation, the lifestyle of my choice.

West Nyack, N.Y.   Steve Goldman

Table of Contents

Preface...v
Acknowledgment...ix
Introduction xvii

Chapter One: The Basics....1
    The Sine Curve 1
    Terminology Describing The Sine Curve....3
    The Application Of Sine Waves To Vibration....5
    Logarithms And Decibels....9
    All-Pass Filtering....11
    Filters.....12
    Three Simple Filters....12
    The Rolloff And Bandwith Of A Filter....14
      The Constant Percent Filter....15
      The Constant Frequency Width Filter....17
    Using Filters For Diagnosis....17
    A Simple Frequency Analysis....20
    Acceptable Levels Of Vibration....21
    The Application Of Sine Wave To Sound Analysis
    Summary....24

Chapter Two: The Fast Fourier Transform Spectrum Analyzer-
              How It Works....26
    Introduction....26
    Why A Spectrum Analyzer....26
    Operation Of A Spectrum Analyzer....27
      Input....27
      Antialiasing....28
      Digitization....29
      The Buffer Memory....31
      Weighting....32
      Fft....34
      Averaging....34
      After The Averaging....35
    Special Sampling Techniques....36
      Order Tracking....36
      Synchronous Time Averaging....40
      Zoom Or Frequency Expansion....41
    The 2-Channel Analyzer....42
    How Big Is A Time Window?....43
    Real-Time Bandwith And Overlapped Processing....44
      Real-Time Bandwith....44
      Overlapped Processing....45
    Dynamic Range:The Big Lie....46
    Summary....47

xii *Contents*

Chapter Three: Transducers For Vibration Measurement....49
    What Parameter To Measure....49
    Motion Transducer Comparison....51
        Proximeters....51
        Seismic Velocity Pickups....54
        Accelerometers....55
    Signal Integration....57
        How To Obtain Velocity-Based Vibration Data....58
        Digital Integration....59
        Analog Integration....60
    The Problem Of Dynamic Range....60
    Mounting Precautions....61
    Force Transducers....63
    Microphones........63
    Hydrophones And Pressure Transducers....65
    Transducer Specifications....66
    Calibration A Spectrum Analyzer For A Specific Transducer....67
    Summary....70

Chapter Four: Elementary Problem Diagnosis....78
    Where To Measure Vibration....78
    Trouble Shooting Guide....79
        Background Readings....79
        Piping....80
        Bases And Support....81
    Single Channel Spectrum Analysis....83
    Forcing Frequencies....83
        Rotational Speed....84
        Flexible Couplings....85
        Blade Frequencies....86
        Gears....87
        Rolling Element Bearings....90
        Fluid-Film Bearings....93
        Cavitation....94
        Looseness And Rubs....94
        Motors....95
    Summary....96

Chapter Five: Dual-Channel Spectrum Analysis....98
    Natural Frequencies....99
        Structural Response Of A Simple Vibrator....100
        The Importance Of Structural Response: An Example....102
        Effect Of Damping On Structural Resonance....102
        The Equations Of Motion Of A Single-Degree-Of-Freedom System....104
        Estimating The Natural Frequencies Of A System....107
        Measuring The Natural Frequency....107
    Cross-Channel Properties....108
        Transfer Function-Mathematical Treatment....108
        Transfer Function-Logical Treatment....110
        Other Kinds Of Inputs For Transfer-Function Testing....116
        Coherence Explained....117
        Some Uses Of Coherence....119
        Coherent Output Power....121
    Modal Analysis....122
    Summary....124

Contents xiii

Chapter Six: Periodic Condition Monitoring....128
   Introduction....128
   The Goals Of Predictive Maintenance....129
   Relevant Terminology....130
      Continuous Monitoring....130
      Periodic Monitoring....313
      Trending....-133
      Untrendable Failures....133
      Diagnostics....134
   Decisions To Be Made When Getting Started....135
      Which Machines Should Be Monitored?....135
      Who Should Decide Which Machines Should Be Monitored?....137
      How Often Should A Machine Be Monitored....138
      What Non-Vibration Parameters Should Be Monitored? 138
      How Are Baseline Criteria Chosen?....139
      Under What Operating Conditions Should Readings Be Taken?....141
      Where Should The Vibration Readings Be Taken?....141
      What Kind Of Vibration Transducers Should Be Used? 142
      What Training Is Required For A Successful Monitoring Program?....143
   A Vibration Monitoring Application Matrix....144
      Logistics....145
      Criticality....147
      Ignore, Replace At Failure-....148
      Overall Continuous Monitoring....148
      Periodic Filtered Monitoring....148
      Overall Continuous Monitoring With Filtered Data Collection....149
      Call A Consultant....150
   What Next?....150

Chapter Seven: Hardware And Software....151
   Introduction....151
   The Vibration Monitoring Device....152
      Octave And 1/3 Octave Band Meters....152
      Tunable Filter Meters....153
      Fft Spectrum Analyzers For Monitoring....154
      Handheld Fft Devices....155
   Software For Trending....160
   Spectrum Analyzers....164
   Recording Devices....165
      Analog Tape Recorders....166
      Digital Tape Recorders....167
   Summary....168

Chapter Eight: Advanced Analyzer Functions....169
   Introduction....169
   Complex Numbers....169
   Input Time 172
   Instantaneous Spectrum....174
   Power Spectrum....175
   Power Spectral Density....176
   Average Cross Spectrum....178

xiv Contents

Transmissibility....179
Transfer Function....179
Force And Response Windows....183
  Input Force Window....183
  Response Window....184
Coherence....184
Coherent Output Power....186
Impulse Response....186
Inverse Transfer Function....189
Summary....191

Chapter Nine: More Advanced Analyzer Functions....192
Introduction....192
Autocorrelation....192
Cross-Correlation....196
Probability Density/Cumulative Distribution Functions....199
Acoustic Intensity....201
Cepstrum....203
Summary....205

Appendix A: Reading Spectral Plots....207

Appendix B: Pulse Theory....211
The Basis Characteristics Of The Pulse....211
Impulsive Testing For Natural Frequency Determination....212
Hammer Hardness Versus Coherence....214
Cavitation And The Pulse....216
Rolling Element Bearing Checkers....218
The Walls Of Jericho....220
Something To Think About....222

Appendix C: Torsional Vibration....223
Introduction 223
The Basics....224
The Advantage Of Torsional-Vibration Readings....225
  Reciprocating Machinery....225
  Sidebands....226
  Flexible Couplings....227
  Gears....227
  Low-Speed Rotation....228
Torsional-Vibration Transducers....229
  Strain Gauges.... 229
  Gear/Sensor Pulse Demodulation....230
  Optical Transducers....230
  Shaft Position Encoders....231
  The Hoodwin Torsional Accelerometer....231
Summary....232

Appendix D: Condition Monitoring Of Reciprocating Equipment....233
Introduction....233
Why Not Use Frequency Domain Data?....234
Pressure As A Machinery Health Indicator....234
Generating The P-V Diagram....235
The Tests....236
Summary....241

Contents xv

Appendix E: Balancing....243
    Introduction....243
    The Causes Of Unbalance....244
    Factors To Consider When Balancing....245
    The Philosophical Problem With Balancing....246
    Vectors....248
    Single Plane Balancing....249
    Multiplane Balancing....252
    Rigid Or Flexible Rotors....253
    Summary....254

Appendix F: Paper Machine Speed-Ups....255
    Advanced Vibration Techniques....256
    Devising A Test....262
    The Test Procedure....263
    Summary....268

Appendix G: Motor And Generator Vibration....269
    Introduction....269
    Ac Motors And Generators....270
    Twice Line Frequency Vibration....270
    Slot Frequency Vibration Of Induction Motors....276
    Stator Core Resonant Frequencies....279
    Rotor Winding Effects On Slot Frequency Vibration And Noise....282
    Slot Frequency Vibration Of Salient Pole Synchronous Machines....282
    Influence Of Air Gap Dissymmetries....285
    Influence Of MMF. Dissymmetries....285
    Summary....286

Appendix H: Data Collector, Spectrum Analyzer Operational Verification Procedures....287
    Required Test Equipment....288
    Initial Setup And Amplitude Accuracy....-290
    Frequency Accuracy....291
    Antialiasing Filter Passband Flatness....292
    ICP Voltage, Current Test And Accelerometer Calibration....292
    Dynamic Range, Amplitude Linearity, Spurious And Harmonic
        Components....294
    Antialiasing Filter Stopband Attebuation....295
    Transfer Function, Gain And Phase Accuracy....295
    Summary....296

Appendix I: Oil Analysis....298
    A Little History....298
    The Three Facets Of Oil Analysis....300
    Machine Wear Analysis....300
    Analytical/Diagnostic Tests....302
    Lubricant Condition Monitoring....304
    Contamination Control....306
    Sampling....308
    Using Your Senses, Post Mortems, And Quality....309
    Equipment Applications....310
    Steps To Good Program Management....316

Afterward....320
Additional Questions....322
Index....329

Introduction

The first edition of *Vibration Spectrum Analysis* was received rather good-naturedly—at least by those readers who, at one time or another, provided some feedback. Both engineers and technicians have found the book easy to read, of sufficient detail to allow anyone to extract practical solutions to their real-world problems from it, and (for a technical book) entertaining.

A quick scanning of the book will take a few hours. Readers can reserve a detailed reading of the necessary bits and pieces required to solve their particular problems as they arise.

In this second edition, the section on how to read spectral plots has become Appendix A. This is because, as the Introduction, it scared people away from reading the rest of the book. Refer to it as you will.

Most of the "meat" of the book remains unchanged, with the exception of a few extra appendices. Questions have been added to the back of the book to prepare the reader for the new "vibration specialist" tests given by some organizations.

Occasionally, short case studies have been added to the back of a chapter to add some additional flavor to the subject. These case studies have been intentionally kept short and vague in order to:

1. protect the identity of the client
2. force the readers to use their imagination so that the problem described can easily be generalized to problems experienced by the readers themselves.

Finally, a bit of advise: Vibration problem solving is fun. Don't ruin it by taking it—or yourself—so seriously that you constrict your thought processes, inhibiting your ability to find imaginative solutions. Let your mind run free, avoiding a "cookbook" mentality. You will do a better job for the equipment under your care.

CHAPTER ONE

The Basics

Introduction

Every rotating machine exhibits a characteristic vibration signature that is uniquely its own. The signal is the sum total of the design, manufacture, application, and wear of each of its components. If the maintenance mechanic or engineer responsible for a piece of machinery takes the time to become acquainted with the nature of the vibration of his machinery, it will not be long before he can effect substantial cost savings for his company.

This chapter originally was written to provide a "no math" way to train servicemen for a leading rotating machine manufacturer. It should go a long way toward teaching the maintenance mechanic how to identify which component of a machine, if any, is about to cause a problem and to estimate the present severity of the problem without opening the machine for inspection. This ability is valuable because it allows for the following:

- Reduction of unscheduled down time
- Reduction of turnaround time
- Elimination of periodic disassembly of a machine for the purpose of inspection
- Greatly reduced probability of a machine "crash"

The Sine Curve

Vibration is defined as a small oscillation about some equilibrium point. To describe vibration, one must speak of the amplitude, frequency, and phase of a series of wave forms called sine waves.

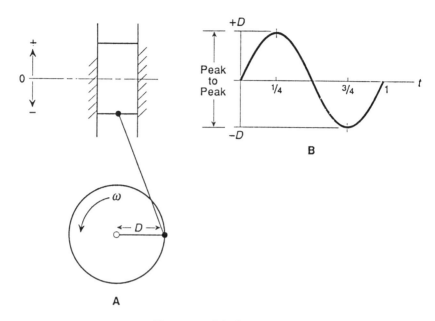

Figure 1.1. Displacement.

Suppose we have a piston driven by a crankshaft moving up and down in a cylinder as in Figure 1.1. The crankshaft is rotating at a speed of one revolution per second (1 Hz). One way to describe the motion of the piston would be to define some point of reference (which we will take as the horizontal center line of the cylinder, or the equilibrium position). When the center line of the piston is above the equilibrium position it is termed positive; when the position of the center line of the piston is below the equilibrium position it is termed negative. Obviously, the greatest distance between the piston center line and the cylinder center line occurs when the piston is at top dead center (positive) or when the piston is at bottom dead center (negative). The center line of the piston will be at the reference point (zero point) when the throw of the crankshaft is at either of the horizontal positions.

We can now plot the position of the piston over time on a graph. When the piston is in the position shown in Figure 1.1A, the amplitude of the deflection is zero, since the piston center line corresponds to the equilibrium position. If the crankshaft is rotating in the direction shown, the piston will be at top dead center in 0.25 of a revolution. Since we are rotating the crankshaft at 1 Hz, this maximum amplitude, which corresponds to the throw of the crankshaft (labeled *D*), will occur at a time $t = 0.25$ sec. As the crankshaft continues to rotate, the piston will move down and in another 0.25 sec will be back in the zero position. This is shown on the curve as zero displacement at time $t = 0.5$ sec. When the crankshaft rotates another quarter turn, the piston is obviously at bottom dead center. This is shown in Figure 1.1B at $t = 0.75$ sec. Finally, in the last quarter of the revolution, the piston moves up and is back to the zero point at time $t = 1$ sec. The plot that results in Figure 1.1 B is called a sine curve and shows two items of interest:

1. The maximum displacement in either direction is D
2. One cycle has occurred in 1 sec

Terminology Describing the Sine Curve

Several things are needed to describe the sine curve in Figure 1.1B. First, the sine curve has a period of 1 sec. This means that it took 1 sec for the piston to start out from the zero point heading upward and again come to the zero point heading upward after one complete cycle. We can also describe this by saying that the piston has a frequency of 1 cycle/sec or 1 Hz.

For the purposes of this book, the units of frequency will almost always be in terms of Hertz (Hz)—cycles per second. Some machinery people insist on describing frequency in terms of CPM: cycles per minute. This has the advantage of relating directly to machine shaft speed in that 1800 RPM (revolutions per minute), for instance, is the same as 1800 CPM (cycles per minute). The major disadvantage is that higher order phenomena, such as a gear mesh frequency, must be described in excessively large numbers. While an 1800 CPM motor driving a 120 tooth gear has a mesh frequency of 216,000 CPM, a 30 Hz motor driving a 120 tooth gear has a mesh frequency of only 3.6 kHz. One should not have to think in terms of numbers larger than one's annual salary.

There are several ways to talk about the maximum deflection of the curve. We can say that the motion has a peak (P) amplitude of D in., making the assumption that, since the curve is symmetrical, there will also be a motion of $-D$ in. in the other direction. We can also say that the curve has a peak-to-peak (P–P) displacement of $2D$, that is D inches up plus D inches down.

The third way to describe the amplitude is a little more complex and is called the root mean square (RMS) value. An explanation of this is as follows: The energy residing in the system described by the

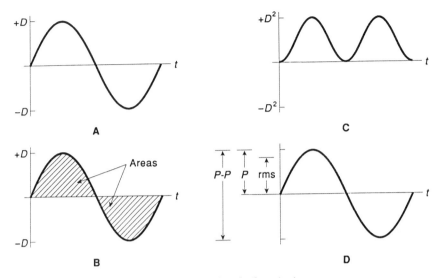

Figure 1.2. Amplitude description.

curve in Figure 1.2A is related to the area under the curve as shown in Figure 1.2B. Some of the area is positive since it is above the zero line and some of it is negative. The problem with this is that, if we attempted to add these areas, we would come out with zero. To get around this, we can square every point on the curve (that is, multiply every point by itself) and plot a curve as shown in Figure 1.2C. Both areas are now positive and we can add them together. To average this value over time, we divide by the period of 1 sec and we take the square root of the answer to come out in units of inches instead of inches squared. This is the RMS amplitude. It can be shown that, for sine waves, the RMS value always equals 0.707 p, where p is the peak value. This is shown in Figure 1.2D. Since there is a 29.3% difference in the value of peak displacement from the value of RMS displacement, it is very important to know what kind of values are showing on a meter. Other ramifications of the difference between using RMS as opposed to peak displacement will be discussed later in Chapter 3. Suffice it to say that almost all meters, from simple voltmeters to multichannel spectrum analyzers, work with RMS values of amplitude regardless of the units in which the instrument is calibrated.

Before continuing, we will give a brief description of phase. Figure 1.3 shows two sine curves, one curve is shifted 90 degrees from the other (just as one piston's motion is shifted 90 degrees from the other in a 4-cylinder engine). It can be said that curve B lags behind

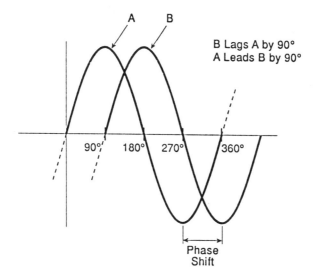

Figure 1.3. Phase shifts.

curve A by 90° or curve A leads curve B by 90°. In Figure 1.3, curves A and B can be said to be 90° out of phase. If we were looking at a piston, as in Figure 1.1, which is rotating at 1 Hz, 90° corresponds to the motion made in 0.25 sec. Therefore, Figure 1.3 shows that curve B is 0.25 sec behind curve A. This is all that we need to know about phase at this time.

The Application of Sine Waves to Vibration

Any vibration occurring at a given frequency can be described by a single sine wave. For instance, instead of describing the crankshaft/piston arrangement of Figure 1.1, we could just as well have described the spring mass system of Figure 1.4, in which the zero point is taken to be at the center line of the mass in its normal resting position. If the mass is now disturbed, it will shake up and down on the spring and describe a motion with time as in Figure 1.4B. In fact, if the mass is oscillating with a frequency of 1 Hz and a peak amplitude of D, the curve will look exactly as it does in Figure 1.1B of the crankshaft/piston example. If, on the other hand, the mass is shaking at frequency of 2 Hz, the curve will appear as in Figure 1.4C. This curve still has peak amplitudes of +/−D. However, this system is operating at 2 Hz, and it therefore makes two complete cycles in 1 sec. The period for this sine curve is only 0.5 sec. From this we can deduce that

$$\text{Period} = \frac{1}{\text{frequency}} = \frac{1}{2 \text{ Hz}} = 0.5 \text{ sec}$$

There are other ways to describe the peak amplitude besides displacement. Instead of measuring the displacement of the piston, we can look at its velocity (see Figure 1.5A). At the top of the piston stroke, the piston has to stop and change direction. This is also true at the bottom, so that it can be said that the velocity in these two crank positions is zero. At the position shown in Figure 1.5A, the center line of the piston is passing through the center line of the cylinder. At this point, the piston is going at its maximum velocity in a positive direction (up). The piston then reaches the top and stops, so that its velocity is equal to zero. As the piston moves down, its velocity increases, and, when the center line of the piston again crosses the center line of the cylinder, the velocity is a maximum in the negative direction (down). The piston will then begin to slow, and, when it reaches bottom, the velocity will be zero because the piston has stopped in preparation for its next upward swing. This can be plotted as shown in Figure 1.5B. Note that at time $t = 0$ the piston is in the position shown in Figure 1.5A and has a maximum upward velocity. At time $t = 0.25$ sec, the piston is at the top and stops (zero velocity). At time $t = 0.5$ sec, the piston is at the zero displacement point again, but moving down with maximum velocity.

The sine curve shown in Figure 1.5B is actually called a cosine curve. If we overlay the curve of Figure 1.5B with the curve of Figure 1.1B, as shown in Figure 1.6, it can be seen that the velocity curve leads the displacement curve by 90°.

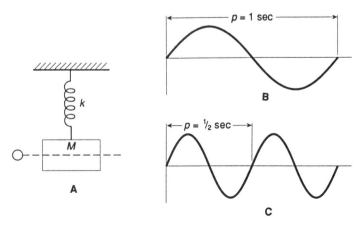

Figure 1.4. Mass-spring system.

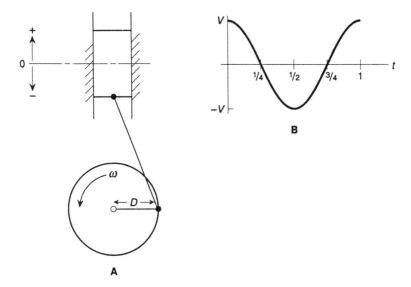

Figure 1.5. Velocity.

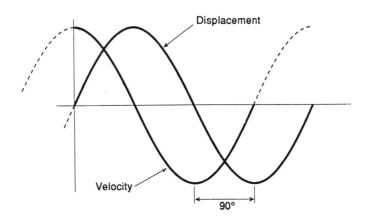

Figure 1.6. Velocity leads displacement by 90°.

Still another way to describe the amplitude of vibration is through the use of acceleration levels. In the piston example of Figure 1.7A, one can see that the maximum acceleration levels occur at the top and at the bottom. At these positions, the piston must decelerate to a stop and then accelerate in either a downward or an upward direction. To put it another way, the largest force exerted on the connecting rod

8   Chapter One

is the force necessary to change the direction of the piston. This occurs with a downward force at the top and an upward force at the bottom. Since $F = MA$ where F represents force, M represents mass, and A represents acceleration, it can be seen that the acceleration is a maximum negative value at the top and a maximum positive value at the bottom. At the point where the center line of the piston coincides with the center line of the cylinder, there is no change in velocity. You would not have to put any force on the connecting rod to get through this point. Therefore, the acceleration at this point is zero. A plot of the sine curve associated with this acceleration is shown in Figure 1.7B. Note that this sine curve starts downward. It is, therefore, 180° out of phase with the displacement curve and 90° out of phase with the velocity curve as shown in Figure 1.8.

Since accelerometers are very inexpensive and easy to use, the measure of amplitude for vibration problems will often be in units of acceleration such as g's or dB of acceleration. The use of decibels to describe the amplitude of acceleration, as well as to describe the amplitude of other quantities is discussed below. The term g of acceleration is equal to 386 inches/sec/sec and is covered in more detail in Chapter 3.

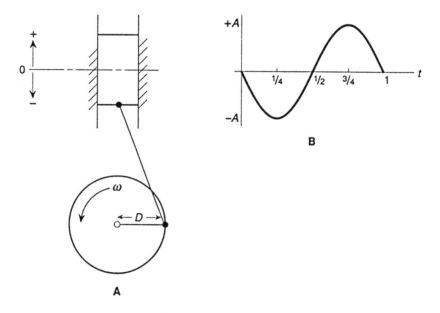

Figure 1.7. Acceleration.

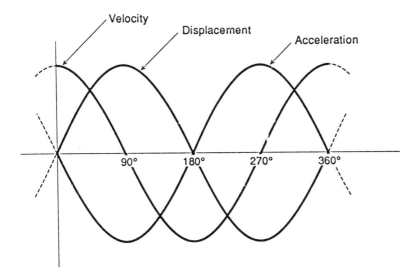

Figure 1.8. Acceleration leads velocity and displacement.

Logarithms and Decibels

On a plot (drawn to scale) showing numbers from 1 to 1,000 on a single line, the numbers would be extremely crowded (as shown in Figure 1.9). For this reason, a mathematical device called a logarithm is used. A logarithm allows you to show numbers from 1 to 1,000 in a manner such that the distance between 1 and 10 is the same as the distance between 10 and 100 and between 100 and 1,000. The advantage is that small values can be seen in the presence of large ones. It is not generally necessary to have as accurate a scale between, say, 1,000 and 10,000 as you have between 1 and 10 (because an error of one unit in the 1,000 range is an error of 0.1%, whereas an error of one unit in the 10 range is a 10% error). Since both vibration and sound measurements go from extremely low levels to extremely high

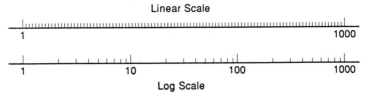

Figure 1.9. Linear and log scales.

levels, a logarithmic scale is often a logical choice for the representation of the amplitude of the motion or sound. To this end, the decibel has been devised. The decibel is defined as

$$\text{dB} = 20 \log \frac{\text{amplitude}}{\text{reference amplitude}}$$

It is necessary to have a reference amplitude because of the nature of logarithms. For vibration, the standard reference is

$$10^{-3} \text{ cm/sec}^2 \text{ for acceleration}$$

$$10^{-6} \text{ cm/sec for velocity}$$

$$10^{-9} \text{ cm for displacement}$$

For sound the standard reference is a pressure of 0.0002 μbar.

By the definition of a decibel, a doubling of amplitude is equal to an increase of 6 dB. Therefore, two machines that generate 100 dB of noise at the same frequency would, if operated in phase, generate a maximum total noise of 106 dB. Table 1.1 gives some other useful values of vibration decibel amplitudes according to the definition above.

Table 1.1. Vibration decibel amplitudes.

dB increase	Linear Multiplication
6	× 2
10	× 3
20	× 10
30	× 30
40	× 100
50	× 300
60	× 1000
70	× 3000

It should be quite evident that no one having 10 fingers would ever have defined a property that has a 20 in its definition. The original concept of a decibel, devised at Bell Laboratories to describe electrical power losses in transmission cables was

$$\text{dB} = 10 \log \frac{\text{power}}{\text{reference power}}$$

The equation for electrical power, P is $P = I \times E$, where I is the current in amperes and E is the voltage. By Ohms Law, $E = I \times R$, where R is the resistance of the circuit in ohms. Substituting E for I in the power equation yields $P = I^2R$.

Therefore, the equation for decibels of power loss in terms of the current amplitude, I, becomes

$$\begin{aligned} db &= 10 \log (I^2R)/(I^2R)_{ref} \\ &= 10 \log (I/I_{ref})^2(R/R_{ref}) \\ &= 2 \times 10 \log (IR)/(IR)_{ref} \\ &= 20 \log (IR)/(IR)_{ref} \\ &= 20 \log (I)/(I)_{ref} \end{aligned}$$

due to the odd fact that

$$\log X^n = n \log X$$

Since, in the study of vibration, we tend to measure amplitude values in terms of peak, peak-to-peak, or RMS levels rather than squared values of power, the equation to use has a 20 rather than a 10 in its definition. This is a problem when speaking to electronic engineers, who use a 10 for power loss in cables, or acoustics people, who use a 10 for sound power readings and a 20 for sound pressure readings.

All-Pass Filtering

If there were no filtering at all, a meter used for the measurement of sound or vibration would add up all of the energy peaks that exist in its sensitivity range (for instance, 25 Hz–8 kHz) and present that total energy as a single amplitude number in decibels or inches per second. A device that only yields a single value of amplitude is very inexpensive because there are no filters involved. Although the single number cannot be used for any diagnostics, it is commonly used as an indicator of the overall severity of sound or vibration existent in the spectra.

Such unfiltered readings are commonly used extensively all over the world to conduct periodic monitoring programs on critical machinery. Unfortunately, the information they convey is woefully inadequate. It becomes impossible, for instance, to see a deteriorating bearing in the presence of a perfectly good gear. (More will be said of this in Chapter 6 on monitoring.) It is therefore necessary to work with data which has first been filtered into useful frequency bands.

Filters

The use of a filter is analogous to the use of blinders on a horse. You put blinders on a horse so that the horse is not distracted by events which are occurring to the left or right, but only sees events straight ahead. In a like manner, a filter can be used to observe only those frequencies that are of immediate interest. Other frequencies are removed from consideration. If the filter is shifted to other frequencies, a frequency-by-frequency description of the vibration signal could be generated.

It is possible to calculate certain forcing frequencies based on the characteristics of the machine. By using filters to isolate the calculated frequencies and by observing the levels of each of them, one can determine whether or not the levels at the forcing frequencies are too high. For example, if a motor is running at 3,600 rpm (60 Hz), one could tell whether or not there was a problem with the balance of the motor by looking at the level at 60 Hz. This would not be possible if one were looking at all frequencies since spikes of higher levels than 60 Hz could exist and it would be impossible to separate them to make sense of the 60-Hz component.

Three Simple Filters

It is advantageous, at this point, to think about three simple types of filters. They are generically called the low pass filter, the high pass filter, and the band pass filter.

The low pass filter only allows low-frequency information to pass through it unattenuated (meaning without a reduction in its amplitude). A 100-Hz low pass filter, for instance, will allow any signal in the frequency range 0–100 Hz to pass through it unchanged. A 200-Hz signal at the input of the filter will be greatly attenuated (reduced) in amplitude by the time it passes to the output of the filter (see Figure 1.10).

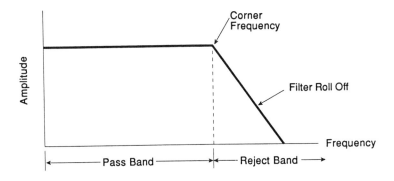

Figure 1.10. A low pass filter.

The high pass filter will pass only those frequencies that exceed the nominal cutoff of the filter without significant attenuation. A 100-Hz high pass filter, for instance, will pass any signal over 100 Hz without attenuation. A signal under 100 Hz will suffer an attenuation of amplitude. Figure 1.11 shows the characteristic shape of a high pass filter.

A band pass filter allows only those frequencies between some lower value and some upper value to pass through it unchanged. In Figure 1.12, it can be seen that a band pass filter looks like the combination of a low pass and a high pass filter.

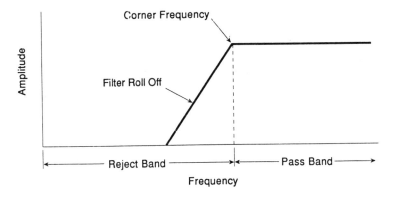

Figure 1.11. A high pass filter.

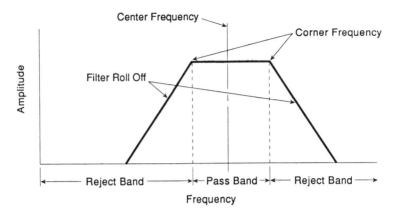

Figure 1.12. A band pass filter.

The Rolloff and Bandwidth of a Filter

Unfortunately, real filters cannot completely chop off a signal that exceeds its nominal limits. In the physical world, only a gradual reduction of energy can take place as the frequency of the signal gets further away from the nominal cutoff of the filter. By definition, the nominal frequency limit of a filter is called the corner frequency. It is that frequency for which an incoming signal will lose 50% (–3db power) of its power before reaching the output terminals of the filter. How severely attenuated a signal outside of the frequency band of the filter becomes is a function of the rolloff of the filter. This parameter is usually stated in dB/octave. Note that an octave is a doubling of frequency, say, from X Hz to $2X$ Hz. Thus, a 6 dB/octave rolloff filter, for instance, would reduce the amplitude of the signal by half for each doubling of frequency beyond the corner frequency.

It would be impossible to see an infinitely thin line on a sheet of paper. A line must have some finite thickness to be seen. For this reason, the 10-in mark on a 1-ft scale does not really show an exact 10-in length, but shows 10 in + / − 1/32 in (depending on the width of the line). In the same way, one cannot see a given frequency without some degree of inaccuracy because this would correspond to an infinitely thin line. Therefore, one must talk about the width of a filter of a certain "bandwidth." Figure 1.13 schematically shows the bandwidth and the center frequency of a filter. The bandwidth is

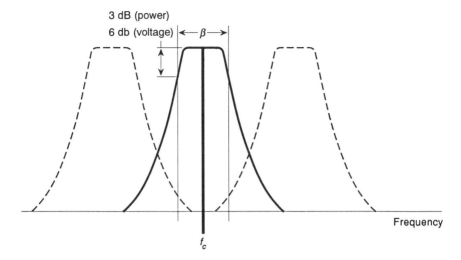

Figure 1.13. Filter bandwidth.

shown as the frequency width between the half-power points (3 dB power = 6 dB amplitude).

It is important to realize that a meter measuring the output of a filter will add up all of the energy in the filter's frequency range and consider all of the peaks to be one peak at the center frequency. Therefore, each center frequency will yield only one number representing all of the energy in the range of the filter. To get an entire spectra (plot of level versus frequency), it is necessary to use either one filter that can be swept through a range of center frequencies or many filters stacked side by side in frequency.

Two kinds of band pass filters are available—the constant percent filter and the constant frequency width filter.

The Constant Percent Filter

The constant percent filter can be tuned to a given center frequency and will see all of the frequencies that exist in the region of + / − some percentage of the center frequency. Most constant percent filters are designed in terms of the octave, for example, 1/3-octave or 1/10-octave bandwidths. An octave is a doubling of frequency. From 2 to 4 Hz is an octave, from 4 to 8 Hz is an octave, from 8 to 16 Hz is an octave, and so on. The problem with constant percent filters is that, as the center frequency gets large, the number of frequencies that can pass through the filter increases and, therefore, the resolution drops. A 1/3-octave filter has three times better resolution than an octave filter but still gets very broad at higher center frequencies.

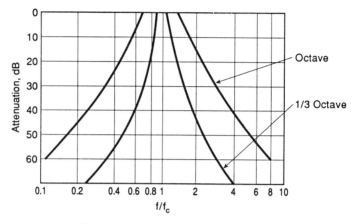

Figure 1.14. Constant percent filters.

Figure 1.14 shows a plot of decibel attenuation (rejection) versus frequency divided by the center frequency of the octave and the 1/3-octave filter. The standard octave and 1/3-octave center frequencies are given in Table 1.2 in Hertz.

Table 1.2. Standard octave and 1/3-octave center frequencies (in Hz).

Octave Bands	1/3-Octave Bands		
31.5	31.5	315	3,150
63	40	400	4,000
125	50	500	5,000
250	63	630	6,300
500	80	800	8,000
1,000	100	1,000	10,000
2,000	125	1,250	12,500
4,000	160	1,600	16,000
8,000	200	2,000	20,000
16,000	250	2,500	25,000

The effective frequency bands for the octave filters can be calculated as follows:

Lower corner frequency = 0.707 × center frequency

Upper corner frequency = 1.414 × center frequency

The effective bands for the 1/3-octave filters, which are 23% wide, can be calculated by substituting the coefficients 0.89 and 1.12 in the appropriate equations above. Other common constant percent filter sets are 1/10 octave (6.9%), 1/12 octave (5.8%), 1/15 octave (4.6%), and 1/30 octave (2.3%), as well as 5%, 10%, and 20% bands.

The Constant Frequency Width Filter

A constant frequency width filter has a fixed bandwidth. If one has a filter with a center frequency of 10 Hz and a bandwidth of 5 Hz, the energy of two signals at, say, 8 Hz and 11 Hz will be added together and considered to be a single peak at 10 Hz. The bandwidth in this case is 50% of the center frequency and may not be as accurate as one might desire. If one were looking at a 5-Hz bandwidth at 100 Hz, the range of frequencies would be from 97.5 to 102.5 Hz. The 5-Hz resolution would, at this center frequency, yield a bandwidth of only 5%. The error becomes even less significant at higher frequencies. If more resolution were needed, it would be possible to go to, say, a 1-Hz bandwidth. For a 10-Hz center frequency, one would see all the peaks that exist between 9.5 and 10.5 Hz and present them as one peak at 10 Hz. Modern spectrum analyzers have the capability to go to bandwidths as narrow as 0.025 Hz or less.

Using Filters for Diagnosis

A comparison of a vibration signal plotted both as octave and 1/3 octave is given in Figure 1.15. Note that a better idea of the characteristics of the vibration is given on the 1/3-octave plot. Also note that much greater accuracy in narrowing down the problem frequencies can be had by looking at the 400 filters (400-line spectra) given in Figure 1.16 for the same vibrating source.

To determine the source of vibration or sound, it is obviously desirable to have the narrowest filters possible. This is because, if one is able to determine the frequencies of the high-level peaks of the vibration signal more accurately, one is more likely to be able to deduce the cause of this high level. The current state of the art allows one to look at a large number of very narrow bandwidth filters. This is a great help in determining what the possible problems of a noisy or vibrating machine are. Table 1.3 lists some of the likely forcing functions and the frequencies at which they can be found. More will be said of this in Chapter 4.

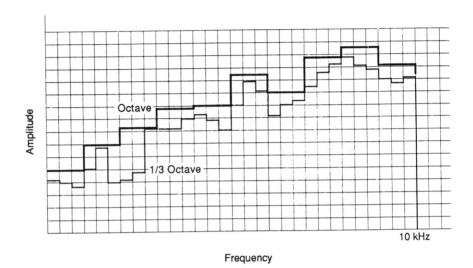

Figure 1.15. Actual octave and 1/3-octave vibration data.

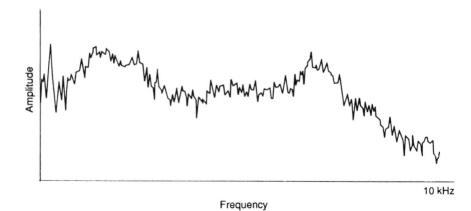

Figure 1.16. A 400-line representation of the data of Figure 1.15.

Table 1.3

Frequency	Cause	Comments
>1/2 × rpm	Oil whirl	Exists in sleeve bearings, change speed, or oil temperature
1 × rpm	Unbalance, eccentric journals	Level is greatest in radial direction Proportional to unbalance
2 × rpm	Mechanical looseness	Look for loose bolts, etc.
1 ×, 2 ×, 3 ×, 4 × rpm	Misalignment of couplings, bearings and bent shaft	Level is greatest in axial direction
1, 2, 3, 4 × rpm of Belts	Bad drive belts	Pulsing amplitude
1, 2, or 3 × frequency of electrical power (usually 50 or 60 Hz)	Electrical ground loops, etc.	Amplitude drops instantly when power turned off
No. of blades × rpm multiples of no. of blades × rpm	Blade frequencies	A high level could mean a stall
Blade frequency $\pm n \times \dfrac{\text{rpm}}{60}$ $n = 1, 2, 3$	Loose coupling	The sidebands are usually quite a bit lower in level than the blade frequency
No. coupling pins, teeth, or grids × rpm	Failing coupling	Could be due to misalignment, wear, lack of lubrication of coupling
1 × rpm of gear No. gear teeth $\times \dfrac{\text{rpm}}{60} \times n$ $n = 1, 2, 3, \ldots$	Eccentric pitch circle, tooth errors in gear Poor gear mesh	Could be due to poor meshing, wear of meshing surfaces, resonance of the gear train. Harmonics could be due to skewed axes of rotation.
Mesh frequency $\pm n \times \dfrac{\text{pinion rpm}}{60}$	Eccentric pinion, pinion shaft not parallel to bull gear shaft, poor pinion support	The more sidebands of high amplitude appear, the greater the severity of the problem
Mesh frequency $\pm n \times \dfrac{\text{bull gear rpm}}{60}$	Eccentric bull gear, bull gear not parallel to pinion shaft, bull gear shaft deflection	See above
$1/2 \, Z f \left[1 + \left(\tfrac{d}{e}\right) \cos \alpha \right]$	Spall, inner bearing race	f = rotational speed, Hz d = ball diameter, in e = pitch diam, in x = contact angle, degrees Z = number of balls
$1/2 \, Z f \left[1 - \left(\tfrac{d}{e}\right) \cos \alpha \right]$	Spall, outer bearing race	See above
$f \left(\tfrac{e}{d}\right) \left[1 - \left(\tfrac{d}{e}\right)^2 \cos^2 \alpha \right]$	Spall, ball of bearing	See above
$1/2 \, f \left[1 - \left(\tfrac{d}{e}\right) \cos \alpha \right]$	Cage unbalance	See above

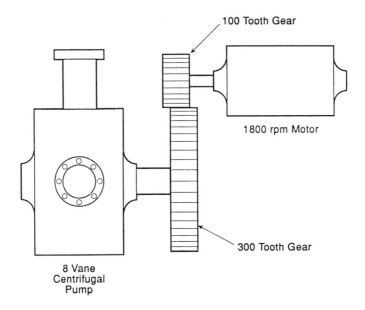

Figure 1.17. Typical mechanical system.

A Simple Frequency Analysis

Consider the system of Figure 1.17. The possible forcing frequencies are as follows:

- Motor frequency = 1,800/60 rpm = 30 Hz
- Pump frequency = 100/300 × 1,800 rpm = 600 rpm = 10 Hz
- Gear mesh frequency = 100 teeth × 1,800 rpm
  = 300 teeth × 600 rpm
  = 180,000 rpm
  = 3,000 Hz
- Blade frequency = 8 vanes × 600 rpm
  = 4,800 rpm
  = 80 Hz

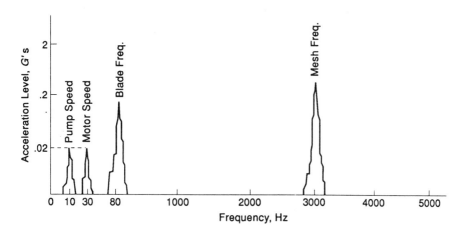

Figure 1.18. Spectrum of typical mechanical system.

Obviously, other frequencies such as gear mesh sidebands, bearing frequencies, and multiples of the above forcing frequencies can exist. A spectrum using assumed values of acceleration levels taken on the pump drive end bearing is shown in Figure 1.18.

Acceptable Levels of Vibration

Having identified the various forcing frequencies in the above example, the next question is whether or not they are at acceptable amplitudes. An acceptable level of vibration does not cause any reduction in the life of the vibrating machine or any damage to nearby equipment and surroundings. Some machinery is designed to tolerate extremely high levels of vibration (such as rock crushers) and some devices are very sensitive to even low levels of vibration (such as optical systems).

There are four ways of determining what level of vibration is acceptable for a given machine. The best way is to keep recorded vibration data at critical locations on the machine periodically over the years. However, since this information is being read now, and not 2 years ago, it will be assumed that this method of determination is not currently available. Once the acceptable levels of vibration are determined a "baseline criteria" can be established. *If you are not already monitoring, start now.*

22 *Chapter One*

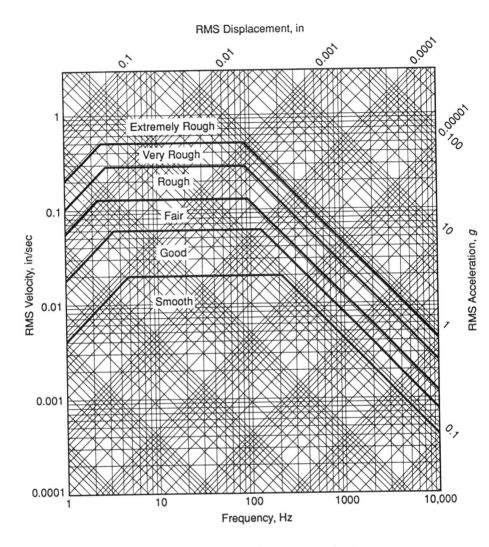

Figure 1.19. Vibration tolerance chart.

A second method of determination is available if a plant has several identical machines. If three machines have similar vibration spectra and the fourth machine exhibits higher levels at the same operating conditions, it is an easy matter to guess which machine is having problems. Another method is to gather vibration data on a suspect machine and mail it to the manufacturer for evaluation. Keep in mind that vibration changes with operating conditions and machine mounting. Also keep in mind that, by this time, you may well know more about vibration than the manufacturer. The fourth method is to choose a standard nomograph based on the experience of others, and, if necessary, modify it to suit your own experience.

The severity nomograph shown in Figure 1.19 has been found to be a useful guide to acceptable levels in typical heavy rotating machinery. Levels of vibration exceeding the "fair to rough" limit should certainly be investigated unless the equipment manufacturer can given assurances that the level is normal for the equipment in question.

Some amplitudes of vibration are unacceptable even at very low levels. This is often true of bearing frequencies, coupling tooth frequencies, and sidebands of +/− shaft speed about a blade frequency, which can indicate a coupling problem. These will be discussed in Chapter 4.

It should also be noted that the gear mesh frequencies of large gear sets do not comply to the severity criteria of Figure 1.19. Experience has shown that the horizontal (constant velocity) lines should be extended from 100 Hz to 10,000 Hz for gear mesh.

The Application of Sine Curves to Sound Analysis

Although the purpose of this book is not to make one an expert in acoustics, the local vibration expert is occasionally called upon to deal with a sound problem. Therefore, some discussion of acoustics will appear at various places throughout this book. The mechanism for sound transport is a series of pressure waves. When the noise source causes a tone, the pressure of the atmosphere just beyond the surface of the source increases somewhat. This high-pressure wave travels through the air and reaches the listener.

The small burst of high-pressure air reacts on the listener's ear as a sound. Between the high-pressure pulses there are regions of air at a lower than atmospheric pressure. Therefore, if pressure over time at the listener's ear was plotted, the curve would resemble a sine curve. It would be exactly a sine curve if the sound generator was a single tone. It makes sense to describe the transmission of noise using sine curves in the same way that sine curves are used to discuss the phenomena of vibration. Sound waves are measured using a microphone, which is really a pressure transducer that is sensitive

24 Chapter One

to small changes in air pressure. It is necessary to use frequency and amplitude to describe sound waves. The stand-ard unit of measurement for the amplitude of a sound wave is the decibel.

Chapter 1 Questions

1. Given the frequency of the pure sine waves below, determine the length of time needed for each of them to complete one full cycle:
   a. 28 Hz
   b. 58 Hz
   c. 120 Hz
   d. 1.2 kHz
   e. 1748 CPM
   f. 3550 CPM

2. Convert each of the following:
   a. 0.73 g's P-P to g's RMS
   b. 7.2 mills P-P to mills peak
   c. 1.2 in/sec peak to in/sec RMS
   d. 2.3 in/sec P-P to in/sec peak
   e. 0.1 in P-P to in RMS
   f. 0.92 g's RMS to g's peak

3. Convert the following frequencies:
   a. 62 Hz to CPM
   b. 3000 CPM to Hz
   c. 127 Hz to rad/sec
   d. 297 rad/sec to Hz
   e. 25 Hz to CPM

4. Convert the following to decibels:
   a. 0.15 in/sec
   b. 8.0 mills
   c. 0.76 g's
   d. 300 in/sec^2
   e. 0.025 in

5. Given the actual pure tone signal at the frequency and amplitude below, determine what the octave band and third octave band amplitudes would be for the filter with the closest standard center frequency.
   a. 60 Hz, 107 db
   b. 522 Hz, 79 db
   c. 1.23 kHz, 87 db
   d. 2.35 kHz, 115 db
   e. 827 Hz, 62 db

Chapter 1 Case Study:

The Problem: GMDC gave a seminar for a refinery in the Northeast. Several weeks later, one of the attendees, a maintenance supervisor, called about a compressor which had been showing high vibration. When asked what the offending frequency was, it was reported that the machine ran at 1770 RPM, and the offending frequency was about 315,000 CPM. The author was asked what could possibly occur 315,000 times per minute.

Solution: The author could not immediately answer the question because 315,000 is a number big enough to be understood only by government accountants. Refusing to be confused by the size of the number, the author divided by 60 to determine the frequency in Hertz: 5250 Hz.

The maintenance supervisor immediately exclaimed that that was exactly what he had calculated as one of the bearing fault frequencies. (see Table 1.3). He decided to change the bearing, and hung up.

Summary

This chapter should have provided enough technical background to allow nonengineers (and those engineers who have not used differential equations for many years) to understand the subsequent chapters. Remember that to do a proper vibration analysis or a good preventive maintenance job, the same two ingredients are required as are required in life—judgement and experience.

Judgement is a prerequisite for being good at anything and it is likely that the reader already has good judgement. The reader may not, however, have much experience. To gain experience in vibration problem solving, simply begin. It is surprising how rapidly one gains experience when doing spectrum analysis. Within a month, the novice has gained considerable ability. Within 6 months, a complete periodic machinery monitoring program can be shown to be cost effective. Within a year, it is likely that the vibration analyst will have committed at least one act of industrial heroism.

CHAPTER TWO

The Fast Fourier Transform Spectrum Analyzer—How It Works

Introduction

A spectrum analyzer is an electronic device that is capable of taking the time waveform of a given signal and converting it into its frequency components. There is an unfortunate irony concerning spectrum analyzers: although they are designed and built by the most sophisticated electronic engineers and computer specialists, their greatest economic value lies in the assistance they give to mechanical engineers and maintenance personnel to keep machinery running. Thus, for years, the machinery man has asked questions in g's, inches per second, and mils and the spectrum analyzer manufacturer has steadfastly insisted on answering in volts, amperes, and bytes.

In this chapter, the operation of a spectrum analyzer will be described in terms familiar to machinery specialists. By knowing how the analyzer works, it is possible to do a more comprehensive job of machinery diagnostics and monitoring.

Why a Spectrum Analyzer

Ever since Baron J. B. Fourier, the French mathematician, showed that it is possible to represent any time waveform (the plot of a signal whose amplitude varies with time) by a series of sines and cosines of particular frequencies and amplitudes, engineers have sought ways to simplify their work by looking for the frequency content of signals. In the machinery world, the goal is to relate the various frequencies seen in a spectra (plot of amplitude versus frequency) to the various physical phenomena occurring in a machine.

For many years, analog filters were used to look at the sound and vibration of a piece of machinery by either having banks of filters side by side and reading the output of each or by using a tunable filter to sweep through the frequency range of interest. There were two major problems: (a) analog filters require long settling times before an accurate amplitude can be measured, and (b) sweeping through a frequency range was so slow that the machine was often under a different load condition at the end of a sweep than it was at the beginning.

In an attempt to speed up the process of gathering data, constant percent filters, such as octave or 1/3 octave, were used. Since a constant percent filter has less and less frequency resolution as frequency increases, an accurate description of machine characteristics was impossible.

In 1965, Cooley and Tukey, of Columbia University, devised a fast Fourier transform (FFT) algorithm, making the calculation of the frequency spectra of a signal far more efficient and rapid than was previous possible in a digital computer.* The FFT spectrum analyzer, capable of rapidly calculating the Fourier transform of a signal, simulated the results that would be obtained by using hundreds of nondrifting, constant-frequency-width analog filters side by side. The spectrum analysis of a sound or vibration signal from a machine could now be easily carried out.

Operation of a Spectrum Analyzer

A properly selected accelerometer (or other transducer) should be used to provide an electrical signal proportional to acceleration to the input BNC jack of an FFT analyzer. The processing of this signal into a valid frequency spectra follows.

Input

The spectrum of the single sine wave A sin ($2\pi ft$) is a single spike of amplitude A and frequency f. If the input voltage to the analyzer is too great for the analyzer to accept, the sine wave will appear to have clipped peaks, somewhat resembling a square wave. If this overload condition is allowed to exist, the analyzer, thinking that it

---

*J. W. Cooley and J. W. Tukey, "An Algorithm for the Machine Calculation of Complex Fourier Series," *Mathematics of Computation*, Volume 19, Number 90, 1965, pp. 297-301.

has some strange waveform to work with rather than a pure sine wave, will try to analyze it. Since the actual signal was a pure sine wave, (distorted only by an improper input attenuator setting) the spectrum of the clipped sinewave is obviously wrong. To avoid this error, set the input attenuator at a level high enough that the overload warning light stays out. Some analyzers do this automatically in the auto-range mode of operation.

If too much attenuation is used, the dynamic range of the analyzer (discussed later in this chapter) will be reduced. Many analyzers have a "low" indicator to warn the operator of this condition. Given a choice between an illuminated "low" light or an illuminated "overload" light, choose the "low" light, as reduced dynamic range is always preferable to a totally distorted signal. Chapter 3 describes ways to deal with this condition.

Antialiasing

Aliasing is the appearance of fraudulent information that occurs because the process of interest was not sampled (observed) often enough. Consider a stagecoach rolling off into the sunset in a cowboy movie. At first, the spokes of the wheels move forward as the coach gathers speed. As the coach further accelerates, the spokes appear to reach a maximum speed, stop, and then, rotate backward–why?

As long as the frame speed of the camera was high compared to the spoke speed, the spokes seemed to accelerate correctly. When the spoke speed exceeded the frame speed, the wheels appeared to move backward. If the horses were fast enough, the coach could go fast enough for the spokes to appear to stop a second time (2 × frame speed) and then begin to accelerate forward. The only way to avoid getting false spoke data in this example is to refuse to observe the wheel during the period that the spoke speed nears the camera's frame speed. An antialiasing filter for cowboy movies would probably be a tachometer-activated lens cap.

A spectrum analyzer is subject to the same kind of phenomena. Since it is a computer-based device, it must digitize the analog signal by sampling it at some frequency (discussed later). The sampling frequency is analogous to the camera frame speed, and the frequency content of the machine to be analyzed is similar to the spoke frequency. The antialiasing filter of a spectrum analyzer is nothing more than a low pass filter that stops any frequency

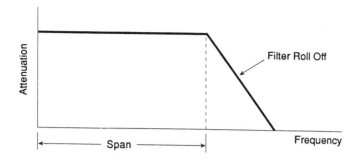

Figure 2.1. The antialiasing filter.

greater than approximately half the sampling frequency from passing through it. Since an actual filter has a rolloff characteristic rather than a discontinuity at the corner frequency as shown in Figure 2.1, the possibility exists that some of the information near the upper end of the frequency span may be contaminated by aliased information. Therefore, many analyzers that have a nominal frequency resolution of 512 lines (or filter bins) actually display only the first 400 lines.

Digitization

Now that we have guaranteed that a fraudulent signal will not be allowed into the analyzer, the analog signal can be converted to digital information in an A/D converter. The signal must be sampled (chopped up) and quantified.

Sampling An FFT analyzer assumes that everything it sees is a series of sine waves. Since everything is a sine wave, the analyzer will need at least two data points to characterize the peak amplitude of each of its 400 filters. Therefore, the sampling frequency must be at least twice the highest frequency of the selected analysis range. If the analysis range chosen is 0–50 kHz, for example, the sampling frequency must be at least 100 kHz (actually, most analyzers sample at a rate of 2.56 × highest frequency to compensate for the rolloff of the antialiasing filter).

Quantification Suppose that one has a partially full bottle of rare wine and four glasses with which to measure the bottle's contents. The glasses are 1 oz, 2 oz, 4 oz, and 8 oz in size. They may be full or empty but cannot be partially full. An example of the measuring procedure is as follows:

- Fill the 8-oz glass
- Pour the excess into the 4-oz glass
- If the 4-oz glass cannot be filled, fill the 2-oz glass instead
- Pour the excess into the 1 oz glass

If the 8-oz, 2-oz, and 1-oz glasses are now full, and the bottle is empty, there were 11 oz of wine in the bottle. The largest amount of wine that can be measured in this four-glass system is 15 oz. If it is impossible to empty the bottle into the four glasses without drinking or spilling some, an overload condition existed.

A 4-bit *A/D* converter has just been described. Note that the accuracy is 1 part in 15 and that the dynamic range is 15/1 or, using the definition for a decibel in Chapter 1, approximately 24 dB. Most spectrum analyzers have a 12-bit *A/D* converter. This means there are 12 binary values used to represent an amplitude (binary values are 0 and 1). An example is given in Figure 2.2. Therefore, the value 13 would be written.

$$000000001101 = 8 + 4 + 1 = 13$$

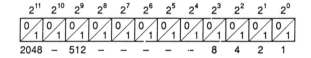

The largest number is:

```
100000000000 = 2048
010000000000 = 1024
001000000000 =  512
000100000000 =  256
000010000000 =  128
000001000000 =   64
000000100000 =   32
000000010000 =   16
000000001000 =    8
000000000100 =    4
000000000010 =    2
000000000001 =    1

111111111111 = 4095
```

Dynamic Range = $\frac{4095}{1}$ = 72 dB

Figure 2.2. A 12-bit A/D converter.

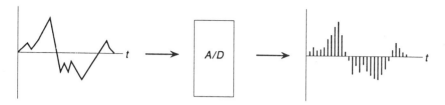

Figure 2.3. A schematic representation of an A/D converter.

The largest value that can be measured with a 12-bit *A/D* converter is

$$2{,}048 + 1{,}024 + 512 + \ldots + 2 + 1 = 4{,}095$$

The dynamic range is

$$4{,}095/1 = 72 \text{ dB } (20 \log 4{,}095/1 = 72 \text{ dB})$$

Figure 2.3 schematically shows an *A/D* converter. It takes the continuous input data from the output of the antialiasing filter and approximates its characteristics by a number of 12-bit digital words.

The Buffer Memory

The *A/D* converter spews out data at the rate of 2.56 times the highest frequency in the range of interest with an accuracy of 1 part in 4,095. Each of these 12-bit words go into a memory called a buffer memory. The buffer memory holds 1,024 digitized time data points for later decomposition to 400 frequency points. Note that 1,024 time data points divided by 2.56 yields 400 frequency points.

Consider the buffer memory to be a carousel that rotates at the sampling frequency and has 1,024 bins, as in Figure 2.4. Each 12-bit word that leaves the *A/D* converter arrives at the carousel at exactly the right time to fill the next available bin.

If one plotted the amplitude value in each successive bin over time in increments of 1/1,024 of the time required to fill the buffer, the result would look like a display of the time domain waveform, as seen on an oscilloscope. Up to this point, we have described a simple digital oscilloscope. When all 1024 bins are full, a "photograph" is taken, freezing the information for use in the FFT calculation. As 12-bit word # 1025 enters the buffer, word #1 falls out, lost forever.

32   Chapter Two

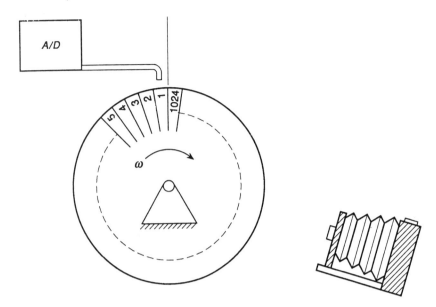

Figure 2.4. The buffer memory.

Weighting

Suppose that a waveform is fed into the carousel such that two succeeding photographs would show a discontinuity (this will happen if the period of the signal does not match the sampling period) as seen in Figure 2.5. The analyzer, thinking that all signals are exactly as seen by the buffer, will assume that the discontinuity occurring between the two sets of time windows is an impulse in the actual data and will present the user with the FFT of a periodic plus an impulse spectra, yielding gross errors. To protect against such false discontinuities, a spectrum analyzer multiplies the time buffer by various windows, which have the property that they multiply the information at the ends of each time window by zero. All adjacent time windows, therefore, will have equal (zero) values at bin 1 and bin 1,024, and the analyzer will not spend time dealing with fictitious impulses. A typical time window is shown in Figure 2.6.

In exchange for avoiding gross errors in a spectra due to fictitious discontinuities, weighting windows may cause slight errors in amplitude values (usually less than 2 dB). This reduction in accuracy is insignificant compared to the large errors that could result from doing no weighting at all.

The Fast Fourier Transform Spectrum Analyzer 33

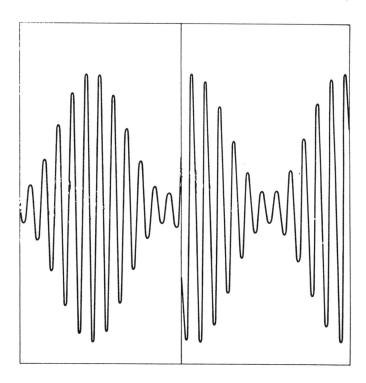

Figure 2.5. Two successive time windows.

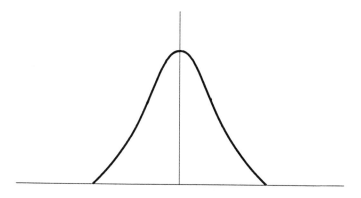

Figure 2.6. A typical time window.

There is one problem with the use of weighting windows. If one were to try to analyze a transient pulse not in the center of time window (they hardly ever are), a weighting function would seriously distort it as in Figure 2.7. For this reason, most spectrum analyzers automatically turn off their weighting functions when in the transient operating mode. This is often called flat, uniform, or rectangular weighting.

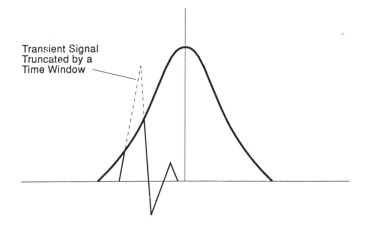

Figure 2.7. Signal distortion due to improper use of a time window.

FFT

Each time the buffer has filled with 1,024 time data points (weighted or unweighted), an FFT calculation is done. This yields, in the case of a 400-line analyzer, magnitude and phase values at each of 400 frequency bins. As mentioned previously, there could have been up to 512 bins, but the upper 112 bins are ignored because of possible contamination from the antialiasing filter rolloff. The resolution (bandwidth) of each of the 400 remaining filters will be

$$\beta = \text{analysis range}/400$$

The amplitude in each bin will be the rms value calculated by the FFT. The phase calculation will be useful only for the 2-channel or synchronous time averaging modes of operation discussed later.

Averaging

There are several types of averaging that can be done by spectrum analyzers. Averaging has the property of enhancing the signal-to-noise ratio of the data.

Summation Suppose one were to take a set of 8 summation averages of a pure tone of, say, 6 volts and 8 random (noise) values of 1, 3, 2, 6, 9, 8, 3, and 5 volts. The pure tone remains pure because

$$(6 + 6 + 6 + 6 + 6 + 6 + 6 + 6)/8 = 6$$

When the noise values are averaged out, the positive numbers settle out to some stable value, in this case

$$(1 + 3 + 2 + 6 + 9 + 8 + 3 + 5)/8 = 4.625$$

All signals are comprised of a steady component (usually the item of interest) and a random component (of no interest). Thus, the method of summation averaging described above will help one "zero in" on the signal of interest with some statistical confidence. The statistical confidence of the readings increases as the square root of the number of averages.

The other common kinds of averaging are as follows:

Exponential The exponential average is a continuously updated summation average in which the most recent data is weighted at 1/(number of averages) of the total information. Eight exponential averages, for example, can be viewed as starting with eight summation averages, removing the first and taking the ninth, then removing the second and taking the tenth, and so on. Each new piece of data counts for 1/8 of the average. It is sometimes useful to watch a run up or coast down of a machine using two or four exponential averages and watching the CRT display for signs of resonances.

Peak This is not really an averaging mode. The analyzer looks at as many instantaneous spectra as desired and saves only the highest amplitude at each frequency bin seen during the data gathering incident. This yields an unaveraged "worst case" result.

Time Instead of averaging sets of 400 pieces of frequency information, the analyzer averages 1,024 pieces of time information. Since the time signal may have negative amplitude values rather than the positive-only values possible from the rms calculation of the frequency domain amplitudes, it is theoretically possible to average the random noise component of the signal to zero. The single FFT of the averaged time trace will have a significantly improved signal-to-noise ratio as compared to the usual summation average.

After The Averaging

The average spectral values calculated above are *D/A* converted (the opposite of *A/D* conversion) for display on a CRT or for plotting on an analog plotter. The digitized information is available in certain formats to feed into a digital plotter, a digital recorder, a

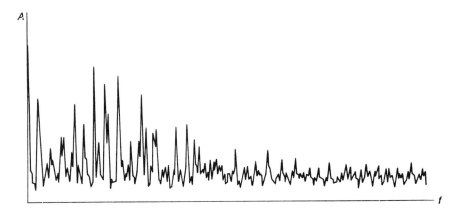

Figure 2.8. A 400-line spectra derived from a 1,024-word time waveform.

computer, or a modem (for telephone transmission to a remote location). An example of such spectra can be found in Figure 2.8.

Special Sampling Techniques

It is possible to greatly enhance your vibration diagnostic ability by using some of the analyzer's capability in special ways. This involves special methods of sampling the data.

Order Tracking

If vibration data is taken on a speed-varying machine, all of the forcing frequencies of the machine will vary with shaft speed. This implies that, as a sum average is taken, each peak will "smear". The peaks will go into different frequency bins, with the result that the actual frequency will be indeterminate and the amplitude will be reduced. For example, consider the mesh frequency and motor pole frequency shifts during 8 averages taken while the speed of the machine varies (see Table 2.1). Assume that the instantaneous amplitude at both the pole frequency and the mesh frequency are 1 g rms (Figure 2.9).

If a frequency span of 5 kHz had been used for analysis ($\beta = 5,000/400 = 12.5$ Hz), the average power spectra display would show a plateau, rather than a spike, from 2,650 Hz to 3 kHz and an amplitude of 1/8 of the correct value. Note that the motor armature frequency, constant at 120 Hz, always registers in the same frequency bin, and will, therefore, yield the correct amplitude. Since the shaft speed peak varies less than one bandwidth, it appears as a single spike.

Table 2.1. Mesh Frequency and Motor Pole Shifts During 8 Averages.

Average No.	Pole	Shaft Speed, Hz	Mesh Freq. (100 Teeth)
1	120	30	3,000
2	120	29.5	2,950
3	120	29	2,900
4	120	28.5	2,850
5	120	28	2,800
6	120	27.5	2,750
7	120	27	2,700
8	120	26.5	2,650

Suppose we took control of the speed of the buffer memory wheel from Figure 2.4. It would no longer be controlled by the internal frequency clock of the analyzer. We could then force it to rotate proportionally with the machine's shaft speed. As the shaft speed increased or decreased, so would the buffer speed. The speed-related information would always turn up in the same bin. A peak at 1 × rpm is called the first order and a peak at 100 × rpm is called the 100th order. The maximum numbers of orders one can see on a spectrum

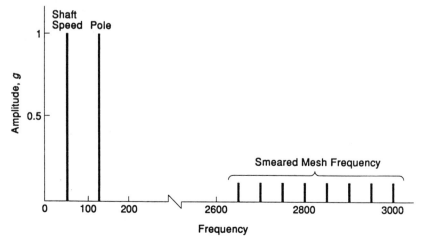

Figure 2.9. Smearing due to speed variation.

38 Chapter Two

Table 2.2. Mesh Speed and Motor Pole Frequency Shifts with 256 Pulses/Shaft or Revolution.

Average No.	Pole	Shaft Speed (order)	Mesh Order (100 Teeth)
1	4.00	1	100
2	4.07	1	100
3	4.14	1	100
4	4.21	1	100
5	4.29	1	100
6	4.36	1	100
7	4.44	1	100
8	4.53	1	100

analyzer is a function of the sampling rate of the analyzer. If the analyzer samples at 2.56 × analysis range, then

$$\text{number of orders} = \frac{\text{pulses/revolution}}{2.56}$$

Thus, to see the mesh frequency of the above mentioned example, the external sampling input of the analyzer would have to be supplied with at least 256 pulses per shaft revolution. If this is done, the example above, in terms of orders, is shown in Table 2.2.

Note that the shaft speed and mesh become narrow peaks with the correct amplitude but that the motor armature varies from 4.00 to 4.53 orders (in this case β = 100 order/400 lines = 0.25 order) and, therefore, smears into several bins as in Figure 2.10.

WARNING. Although most of the analyzer works in orders when in external sampling, the antialiasing filters are fixed in fre-

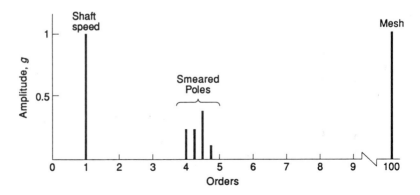

Figure 2.10. An order plot of smearing due to speed variation.

quency. For large-speed variations such as in the run up or coast down of machinery, it is necessary to use a tracking antialiasing filter. Figure 2.11 shows the kind of errors an improper antialiasing figure can cause when order tracking. Figure 2.11A shows the actual signal. In Figure 2.11B, the right side of the plot rolls off to zero, because the antialiasing filter was set too low and actually chops off the higher order data. In Figure 2.11C, there are extra peaks because the antialiasing filter was set too high to prevent aliasing.

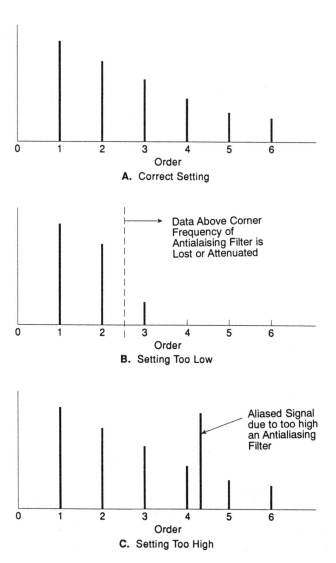

Figure 2.11. *Effect of antialiasing filter setting on order tracked data.*

Synchronous Time Averaging

As mentioned above, time averaging enhances the signal-to-noise ratio. Synchronous time averaging enhances machine diagnostic ability as well. In synchronous time averaging, the input buffer begins taking 1,024 time data words when it receives an external pulse. After the 1,024th word, the analyzer stops and waits for another external pulse before reloading the buffer.

If the external pulse comes from an optical or magnetic pickup looking at a keyway on a shaft, the time data gathered will always start at the same shaft position. If the time signal is averaged over a large number of data blocks, therefore, all events related to shaft position will reenforce themselves. Those signals that are asynchronous with shaft motion will appear to be random in nature and eventually average out to a low value. After the time signal has been averaged out, a single FFT of this time signal is made.

Most older single-channel spectrum analyzers were not capable of time averaging because they had insufficient memory to be able to average sets of 1,024 time points rather than sets of 400 frequency points. Older 2-channel units could only time average in the single-channel mode by reallocating the memory usually reserved for the second channel. Since electronic memory is now quite inexpensive, it is uncommon to find an analyzer that cannot time average.

Three useful examples of synchronous time averaging are given here.

Which Value is Bad? If one mounts an accelerometer on a reciprocating machine, uses a 1/rev signal from the shaft for synchronous time averaging, and chooses a time window equal to or larger than the time required for the machine to make one complete cycle, a time trace showing the tapping of each valve will result. To determine which of the valves is causing a problem, read the time interval from the trigger point to the highest amplitude spike (knowing shaft rpm). The troublesome valve can thus be located.

Which Gear is Bad? A high amplitude at mesh frequency indicates that one of the meshing gears has a problem. Synchronously time average the gearbox vibration using a 1/rev pulse from the keyway of each shaft in the gearbox successively. A spectrum of the average time trace with a high level at mesh frequency indicates that the gear on that shaft is the culprit. You could even predict which tooth is bad by using the bad valve technique described above but it is difficult to replace a single tooth.

Balancing Balancing a rotating mass requires the ability to measure the effect of trial weights on the amplitude and phase of the vibration at running speed. By synchronous time averaging, it is simple to obtain amplitude levels. Also, since it is easy to determine which of the 1024 time history words has the peak amplitude relative to the starting pulse supplied to the analyzer (keyway position), the phase shifts at running speed due to trial weights can be obtained. Given this information, one could balance using the method that is used with an archaic strobe light-type balance machine.

NOTE: It is possible to do synchronous time averaging while order tracking, since they are two unrelated variations to sampling. In order tracking, the speed of the memory wheel is varied, whereas with synchronous time averaging the wheel is started at the appropriate moment and stopped after each full rotation (regardless of the speed of that revolution).

Zoom or Frequency Expansion

Often, it is necessary to have better frequency resolution at relatively high frequencies than can be obtained by dividing the analysis range from 0 Hz to some frequency above the area of interest by 400 lines. This problem is handled by requesting the analyzer to look at some frequency span that will give the desired resolution and centering that span around the area of interest. There are three ways in which this is handled by spectrum analyzers.

Analog Range Translation In this oldest, and least desirable method of zoom, a fixed span, which had to be purchased as an option to the analyzer, could be centered about any frequency of interest by modulating that frequency with some carrier frequency, doing an FFT calculation, and demodulating the results. The analyzers that offered this type of zoom allowed the purchaser to order any two of several possible span windows. The most common pair were the 50- and 500-Hz windows, allowing for frequency resolutions of 0.125- and 1.25 Hz bandwidths, respectively.

Digital Range Translation In this most common, as well as most versatile, method of zoom, the user selects a span and a center frequency for the span. The sampling rate slows down to whatever rate is required to obtain the desired filter bandwidths, and an extra term appears in the FFT calculation to shift the span about the desired center frequency. As in the analog zoom described above, each new span or center-frequency selection requires resampling. The advantage to this type of zoom is that a very large number of spans are available at any center frequency within the operating range of the analyzer. In some modern analyzers, the only true baseband FFT is from 0 to 50,000 or 100,000 Hz, and all other ranges are actually different spans that may or may not begin at 0 Hz. The advantage to this kind of architecture is its cost. Only a very few analog antialiasing filters must be used. Further antialiasing is accomplished with less expensive digitally simulated filters. This allows the purchaser to obtain a spectrum analyzer with integral zoom at less cost than a design with a zoom option costing $3,000 or more.

Large Memory Buffer Translation This newest implementation of zoom has come about because of the significant reduction in cost of computer memory chips. The method is quite simple. Instead of gathering only 1,024 time history points at a time, this method allowes the analyzer to take, say, 10,024 time points. This may take 10 times longer to do in any given analysis range, but the frequency resolution will be 10 times better as well. Any center frequency within the chosen analysis range may be selected and the span will be, in the case of a 10 × zoom, 1/10 the baseband range. For example, if one were to desire to examine a signal at 4,500 Hz, the 0–5,000-Hz baseband would have to be selected. The span would be 500 Hz wide with a 1.25-Hz resolution. If more resolution were needed, an analyzer with a 10 × zoom would be inadequate. One advantage to this type of zoom is that it is possible to zoom in on several different center frequencies simultaneously with the same frequency resolution because the analyzer uses the same antialiasing filter for any center frequency in the range and can use the same 10,000 memory point sampling. Another advantage is cost. This type of zoom is seldom priced as an optional extra, but it is included in the architecture of the unit.

The 2-Channel Analyzer

A 2-channel spectrum analyzer is far more powerful than two single-channel analyzers. This is because, by preservation of the phase relationship between the two channels, cause and effect relationships may be established.

The 2-channel analyzer operates in the same way as a single-channel analyzer with the following exceptions:

- Two input attenuators
- Two input buffers, controlled by the same internal clock or external sampling signal
- Half the number of lines of resolutions as the same analyzer in the single-channel mode (usually)
- Calculation of cross-channel properties such as transfer function, coherence, and coherent-output power

A discussion of the various dual channel functions and their uses will be reserved for future chapters.

How Big is a Time Window?

As mentioned previously, a spectrum analyzer looks at a minimum of two points in the time waveform signal to determine the characteristics of the sine wave passing through these points. The time required to look at these points for the finest resolution sine wave that can be seen by the analyzer for a given analysis range is called the time window, or memory period. This value is equivalent to the time it takes for the memory buffer to fill with 1,024 time points.

Suppose one were to look at a frequency span of 500 Hz with 400-line resolution. The smallest frequency increment that could be seen is 500/400 = 1.25 Hz. The period of a 1.25-Hz sine wave is 1/1.25 = 0.8 sec. Therefore, in this analysis range, the memory buffer must fill its 1,024 bins in 0.8 sec. If the analysis range were increased to 5,000 Hz, the resolution would decrease to 12.5 Hz, and the time window would decrease to 1/12.5 = 0.08 Hz. For 10% of the original resolution, we need only spend 10% of the time observing the signal. Also note that, as the resolution improves (as in zoom), the time required to gather data increases independently of the design of the analyzer or the kind of zoom it employs.

The amount of time required when tape recording a signal for later spectrum analysis can be estimated to be slightly more than the number of averages × time window. The frequency ranges of interest and the number of averages needed for good statistical confidence must, therefore, be decided before the data is tape recorded.

When synchronous time averaging, a frequency range should be chosen such that the time window is at least large enough to cover a full revolution of the shaft to catch any event that may appear during shaft rotation. When order tracking, the time window must be calculated based on the order resolution given, for example, for a 400-line analyzer, as

Order time window = 400 / number of orders

Real-Time Bandwidth and Overlapped Processing

There are two subtle issues that arise from the use of an FFT analyzer which may, on rare occasions, affect the quality of data gathered. They are the topics of real-time bandwidth and overlapped processing

Real-Time Bandwidth

If one thinks about the timing of the effort an analyzer must make to present the user with a completed frequency spectra, the notion of real-time bandwidth becomes a simple one. Think of the memory buffer as a rotating wheel as described above. When the wheel has taken on 1,024 time points, the data is multiplied by the proper weighting window and the FFT algorithm and other operations are performed. This may take, say, 200 msec. During this time, the memory wheel continues to turn, with an old piece of time data falling out each time a new 12-bit word enters one of the 1,024 slots. Once a piece of data leaves the buffer memory, it is lost forever. If this on-going data-replacement process occurs at a slow enough rate, by the time all 1,024 pieces of data are replaced, the analyzer will have completed its number crunching and will be ready to look at the new set of data. If the memory wheel rotates too quickly, the analyzer will not have completed its work from the previous 1,024 points and some new data will have had time to enter the memory wheel, rotate one complete revolution, and fall out before the analyzer had a chance to see it. There will be a hole in the data history.

The real-time bandwidth of a spectrum analyzer is the maximum frequency that can be measured without missing data. If the number crunching takes 200 msec, and the analyzer has 1,024 memory bins and a sampling rate of 2.56 × analysis range, the highest frequency range which can be analyzed without losing some data is 1,024/2.56/0.2 = 2000 Hz or

$$\text{Real-time bandwidth} = \frac{\text{number of lines of resolution}}{\text{processing time}}$$

Some spectrum analyzers, because of their design, are actually catch processors. In such a design, the memory wheel stops after 1,024 points have been gathered and waits for the processing to be completed before gathering more data. Since data is ignored each time the wheel stops, no frequency is low enough to avoid being partially missed, and the real-time bandwidth is zero. In other designs, it is possible to increase the real-time bandwidth somewhat by turning off the display. This is one less activity demanded of the processor, allowing it to return to the memory wheel sooner. Most spectrum analyzers have real-time bandwidths from 2 to 10 kHz. For most machinery applications, changes in the vibration characteristics occur so slowly that the loss of some data makes no difference to the analysis. The following is an easy-to-follow example of a case where an inadequate real-time rate can make a difference.

Suppose one were photographing the animals on a merry-go-round. The horses are arranged such that after each 9 horses, there is a swan for the little children and their mothers to ride in. There are 27 horses and 3 swans. The merry-go-round rotates at 0.5 rpm, having an animal-passing frequency of 15 animals/min. To photograph all of the animals consecutively, you would have to shoot at a real-time rate of 15 frames/min. If your camera could only shoot 1.5 frames per minute, you might well walk away from the merry-go-round thinking that all of the animals were swans, you missed all of the horses.

Overlapped Processing

Overlapped processing is the other side of the coin from the real-time bandwidth problem. Suppose, as above, the number crunching done by the analyzer takes 200 msec and the analysis range chosen is 0–50 Hz. It is desired to take 32 averages of the signal.

The time window is 400/50 = 8 sc. The memory wheel fills in 8 sec and the processing takes 0.2 sec to make the first set of calculations. When the processor returns to the memory wheel only 1,024 × (0.2/8) = 25.6 memory bins have been updated. Since it makes very little sense to take a new average when only 2.5% of the data has changed, the processor will have to wait for some time before taking 1,024 data points for the next average.

If the analyzer waits for the memory wheel to fill with 1,024 new data points, there will be no overlapped processing. The time to take eight averages will be 32 × 8 + .2 = 256.2 sec.

Overlapped processing means that the processor will not wait for a new set of 1,024 time points but will begin processing after only a certain percentage of new points are added. For instance, a 25% overlap means that 768 new points and 256 old points will go into the next FFT calculation. The averaging will be 25% redundant.

Some analyzers never do overlapped processing, some will automatically overlap from 0-50%, depending on the relationship between processing time and time window. It is crucial to know what a particular analyzer will do. It turns out that, while 32 averages at 50% overlap takes about as long as 16 nonoverlapped averages, the confidence limit is only slightly better than 16 nonredundant averages. Thus, if a 90% confidence level of 1dB or better is needed, either 32 nonredundant averages or 64 50%-redundant averages need to be taken. The author knows of a researcher who had to throw away 7 months of data because he did not realize that he was taking 32 50%-redundant averages rather than the 32 nonredundant ones he needed to prove his theory with an adequate confidence limit.

A few analyzers have selectable maximum amounts of overlapping and some allow the user to force the analyzer to average as fast as possible without respect to redundancy. This ability becomes useful in swept sine tests to insure that data is measured at as many discrete frequencies as possible during a sweep.

Dynamic Range: The Big Lie

As mentioned earlier in this chapter, dynamic range is the ratio of the largest amplitude signal that can be seen with a given attenuator setting to the smallest observable signal. It was pointed out that, for a 12 bit A/D converter, the dynamic range is:

$$20 \log (4095/1) = 72 \text{ db}$$

As an exercise in one-upmanship, some spectrum analyzer manufacturers have gone to 16 bit A/D converters and claim a 96 db dynamic range. Often, this is a lie.

That the A/D converter can sense one part in 16 binary bins is no assurance that the analog circuitry is good enough to insure that the information going into the lower bins is not contaminated by electrical noise.

An excellently designed analyzer might be able to muster a respectable 80 or 85 db dynamic range in the analog front end of the device regardless of whether it has a 16 bit A/D converter or a "dithered" 12 bit converter. The author knows of one 12 bit, "72 db" dynamic range analyzer that could show only a 60 db scale on its screen because of circuit noise. Let the buyer beware.

Chapter 2 Questions:

1. Determine the minimum time window to take a single sample of data using a 400 line spectrum analyzer (no Zoom) in the following frequency ranges.
   a. 0–50 Hz
   b. 0–200 Hz
   c. 0–500 Hz
   d. 0–1kHz
   e. 0–5kHz
   f. 0–10kHz

2. For the frequency ranges above, determine the time window for the analyzer set to zoom to a decimation ration of 5 (5x zoom).

3. Assuming a 100 msec. processing time for each instantaneous spectra, how long will it take to process 16 nonoverlapped summation averages of a signal in each of the frequency ranges of question 1? How long for 50% overlap?

4. A machine has major forcing frequencies at 1x, 2x, 20x, 40x, 237x, and 474x RPM. What analysis range would be needed to see all of the peaks if the speed of the machine is:
   a. 10 Hz
   b. 1150 RPM
   c. 1770 RPM
   d. 30 Hz
   e. 3000 RPM

5. For the machine described in question 4, and a spectrum analyzer range of 0–5 kHz, what speed variation can be tolerated without smearing (thus forcing one to order track) in order to see the highest of the forcing frequencies visable in the 5 kHz range for running speeds of:

a. 10 Hz
b. 20 Hz
c. 30 Hz
d. 60 Hz

Chapter 2 Case Study:

The Problem: A major drug company had a tower of several stories height. Within the tower, a chemical dust was allowed to fall a considerable distance to a separator of some kind. Corporate office engineers measured a major tower vibration peak at 10 Hz, but could not find its cause through finite element analysis, and requested GMDC to find the cause of the offending 10 Hz peak.

Test Measurements: After extensive single channel measurements of the tower at various separator operating speeds, and quite a few dual-channel tests to attempt to locate a 10 Hz natural frequency in the structure, nothing in the region of 10 Hz was found, although several high amplitude frequencies—already known to the engineers as being related to the separator—were found.

The Results: There was no 10 Hz vibration signal. The corporate engineers had purchased an inexpensive FFT board for use with their portable computer. As you usually get what you pay for, the FFT board did not have antialiasing filters. The 10 Hz signal was an alias of one of the separator forcing frequencies.

Summary

The spectrum analyzer is an extremely powerful tool for machinery health monitoring and diagnostics. Its proper application requires a rough idea of how an analyzer works. This chapter was written to help the person responsible for machinery operation and reliability to get a feel for the way in which an FFT spectrum analyzer operates. Without this knowledge, the chances of choosing the best piece of equipment to do a particular job is small. Worse, without this knowledge the person responsible for diagnosing problems will be forever constrained to work with only the most simple-minded tools and techniques.

CHAPTER THREE

Transducers for Vibration Measurement

The proper gathering of vibration data for machinery health monitoring and diagnosis relies on the proper selection of transducers. This chapter deals with the three most common motion transducers: proximeters, velocity pickups, and accelerometers, as well as piezoelectric force transducers, microphones, and piezioelectric pressure transducers.

What Parameter to Measure

Since the velocity pickup was the most common vibration transducer available in the early days of predictive maintenance work, the decision as to what parameter should be measured was unconsciously made in favor of velocity (usually in in/sec 0–peak). Some early investigations, attempting to choose the correct parameter, verified that velocity seemed to be a good choice. Certain assumptions about the vibration characteristics of machinery, which were correct only for some classes of equipment, were made. Further thought on this matter, as detailed below, shows that the decision favoring the reading of displacement, velocity, or acceleration levels should be a function of the amplitude values and relative importance of the various forcing frequencies that a machine is capable of generating.

The choice of whether to measure displacement, velocity, or acceleration is not as obvious as one might hope. That a given facility has been measuring displacement or velocity exclusively for many years is a weak excuse for carrying on a tradition that may be incorrect for a certain percentage of the equipment being used in the plant. This section will give the reasons for measuring a particular parameter on a particular machine.

Explanations of the meanings of the terms displacement, velocity, and acceleration were given in an empirical way in Chapter 1. It will now be necessary to look at a little math. Assume, for the moment, that the machine of interest is vibrating at a single frequency (a simplifying, although unlikely assumption). The displacement of the vibration as a function of time will be

$$x = X \sin(2\pi ft)$$

where X is the peak displacement and f is the frequency. The displacement at any instant of time is x.

The velocity equals the difference in displacement divided by the difference in time to go from one position to another, or $v = dx/dt$. In a like manner, acceleration is the rate of change of velocity, or $a = dv/dt$. Through the magic of the study of differential equations (or blind faith for those not interested in differential equations), putting the equation for x above into these new equations for velocity and acceleration yields

$$v = 2\pi fX \cos(2\pi ft)$$

$$a = -(2\pi f)^2 X \sin(2\pi ft)$$

Table 3.1. Comparing Displacement, Velocity and Acceleration Values at Different Frequencies.

	Equation	Propeller Speed	Turbine Misalignment	Turbine Blade
Frequency (Hz)	—	0.83	120	12,000
Displacement (in.)	X	X	X	X
Velocity (in./sec.)	$2\pi fX$	$5.2 X$	$754 X$	$75.4E03 X$
Acceleration (in./sec.^{2})	$2\pi f^2 X$	$27 X$	$569E03 X$	$5.69E09 X$

Because our measurement will actually be in peak or rms units, it is possible to ignore the sine and cosine functions and write the values for maximum amplitude. Table 3.1 gives the equations for these values, as well as what they would be for a system using a 3,600-rpm steam turbine with 200 blades to drive a 50-rpm four-bladed ship's propeller.

For the turbine blade frequency level to have a reasonable acceleration of 0.1 g (38.6 in/sec^2), the displacement would have the unmeasurable value of 6.9E–09 in. This small value of displacement is arrived at by dividing the acceleration value of 38.6 in/sec^2 by the displacement to acceleration correction factor of 5.69E09 as found in Table 3.1. We may infer from this that displacement is not a good measure in the high-frequency range.

A reasonable displacement amplitude at the propeller shaft speed is 0.005 in (an easily measured value). The equivalent acceleration level of 27 times 0.005 = 0.135 in/sec^2 (0.00034 g) is not, however, a reasonable amplitude to deal with. Low-frequency signals do not lend themselves to acceleration measurements.

Velocity readings, obviously, fall in the mid-frequency range. Many people maintain that, for this reason, velocity is a good indicator of machinery condition in most cases. A more proper selection of the amplitude parameter to be used at different frequencies is given in Table 3.2. Figure 3.1 shows a comparison of typical vibration data as seen in terms of acceleration, velocity, and displacement. Note the wide range amplitude scale of 120 dB.

Table 3.2.

Parameter	Displacement	Velocity	Acceleration
Frequency, Hz	0–30	5–2,000	Over 50

Motion Transducer Comparison

Unfortunately, the physics of the operation of the various kinds of transducers are such that they have limited frequency ranges. Simply because one is interested in a frequency range of, say, 5–2,000 Hz does not mean that a velocity pickup is the obvious transducer to use. One must be constantly alert to the many problems that can arise from poor transducer selection.

The transducers most commonly used for vibration are proximeters, seismic velocity pickups, and accelerometers. Their construction and method of operation will be discussed in the following sections and warnings concerning their proper use will be given.

Proximeters

A proximeter is a device that generates an electromagnetic field at its tip. The action of the proximeter tip coming close to a metal target causes interference in the magnetic field, which changes the

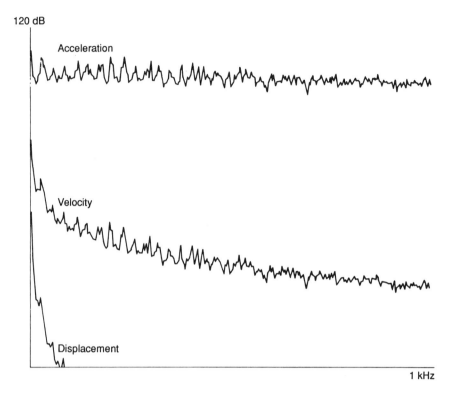

Figure 3.1. A comparison of the same vibration data in terms of acceleration, velocity, and displacement.

output of the unit. Since the output of the probe (also called an eddy current probe) is a nonlinear function of the gap between the probe tip and the target, a linearizing circuit must be used. The output of the linearizing circuit is proportional to gap from direct current (dc) to a few hundred Hertz. Note that the proximeter simply reads gap—it has no idea whether a shaft is rotating or simply shaking back and forth in the plane of the probe (see Figure 3.2).

Advantages Since the proximeter can read dc clearances, they provide a very good way of monitoring such parameters as oil-film thickness in oil-film thrust and radial bearings. Proximeters are a popular way of observing shaft orbits because they measure the gap between a stationary member on which they are mounted and the shaft directly. Thus, if one mounts two proximeters at right angles to each other, such that one probe observes the vertical motion of a shaft and the other the horizontal, the output of the probes can be used to

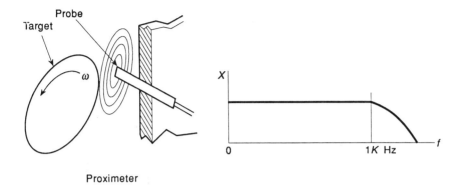

Figure 3.2. The operation and typical frequency response of a proximeter.

drive the vertical and horizontal amplifiers of an oscilloscope to yield shaft orbits.

Proximeters are also useful for making vibration measurements on a low-speed shaft supported by oil-film bearings. This is true because they can read the shaft directly rather than trying to detect what little motion transmits across the oil film to the bearing housing.

Disadvantages Proximeters should only be used to frequencies of a few hundred Hertz. The exclusive use of these devices to monitor vibration will result in the loss of all higher order information (such as information at blade and mesh frequencies). It is a common practice in some installations, in fact, to put low pass filters in orbit-measuring systems, such that anything above the fifth order is filtered out for clarity in viewing low-speed phenomena. Keep in mind that a failure may well result in some higher frequency mechanism to which a proximeter-based monitoring system is blind.

If the shaft material is not magnetically uniform, the proximeter will see a 1 × rpm signal even for a perfectly aligned, zero-vibration shaft. This makes it necessary to gather slow-roll information on a proximeter/shaft arrangement before drawing any conclusions as to machine health. Discontinuities such as scratches and dents on the shaft surface have the same effect. Some modern proximeter systems allow the user to digitally save the slow-roll information and subtract it out of all future readings.

Proximeters read relative motion between the probe and the target. It is therefore necessary to mount the probe in such a way that it does not vibrate itself, but measures only the target vibration.

A proximeter properly arranged to see shaft vibration will totally ignore casing vibration. A turbine with dozens of proximeters looking at the rotating assembly, therefore, can have cracking stator blades, which will be ignored by the vibration monitoring system until one of the stator blades breaks off and passes through the rotor blades. Such a warning system may prove to be of little value.

Seismic-Velocity Pickups

Before the introduction of integrated circuit piezoelectric accelerometers, most vibration technicians used seismic-velocity pickups to measure vibration. This was because, in spite of their poor frequency response, large cross axis effects, and cumbersome size, they were less painful to use than the delicate, temperamental, and expensive charge-amplifier-type accelerometers available at the time.

A seismic-velocity pickup is made up of a magnetic mass on a spring in a bath of oil. Surrounding the mass is a coil. When the pickup is subjected to vibration, the magnetic mass oscillates relative to the coil. In the same way as a generator makes electric power by rotating a magnetic field past a stationary coil, the oscillating magnet causes an electromagnetic field to generate a voltage in the coil. The voltage is proportional to the velocity of the surface on which the pickup is mounted (see Figure 3.3).

Advantages The velocity pickup is sturdy and has been around for many years. People are familiar with its use and are comfortable with its beer-can shape. Since the signal is self-generated, a power supply or charge amplifier is unnecessary.

Disadvantages The equation for the undamped natural frequency of a single-degree-of-freedom system (one mass, one spring), is

$$f_N = 1/(2\pi) \sqrt{K/M}$$

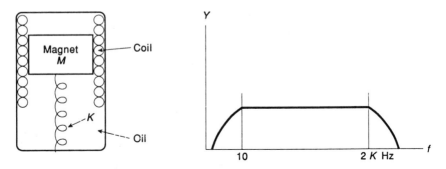

Figure 3.3. A seismic velocity pickup and typical frequency response.

where K is the spring stiffness and M is the mass. In a velocity pickup, the steel spring is relatively weak and the mass is large. Therefore, the natural frequency of the pickup is low—about 10 Hz. To avoid vibration readings that are greatly magnified at 10 Hz due to the resonance, the pickup is filled with oil, which overdamps the transfer function (see Chapter 5), causing the sensitivity of the pickup to roll off in response below 10 Hz. This is the supreme irony: people all over the world are using velocity pickups because they have low-speed machines; however, the pickups themselves become insensitive below 10 Hz!

Further disadvantages to velocity pickups are their sensitivity to cross-axis effects (if the pickup is mounted vertically, it will also see horizontal motion), the fact that some designs can work only in one direction (a unit purchased for vertical mounting may not work in the horizontal direction, and *vice versa*), their frequency response rolls off at about 2 kHz, and they are large enough to cause mass loading on small, light mounting surfaces. Velocity pickups that do not have some of these problems are available but they are quite expensive.

Accelerometers

Accelerometers are seismic instruments. They are made of a very small mass mounted on a piezoelectric crystal. The crystal acts as a stiff spring, as well as having the property of generating a charge proportional to acceleration.

In early accelerometers, the low-energy charge was brought through a cable to a charge amplifier for conversion to a voltage proportional to acceleration. Since the charge itself is extremely small, the output of the accelerometer varied with cable length to the amplifier, kinks in the cable, lying the cable over a sharp metal edge, the swinging of the cable near an electric field (such as a motor), and contamination of the connections in an oily environment. Accelerometers used to self-destruct if dropped because there was no internal electronic protection. All of these problems led to the rise of the velocity pickup.

In the capacitive type accelerometer, the acceleration causes a change in capacitance between the seismic mass and the electrodes. The change in capacitance is converted into a high voltage, low impedance output by the built-in electronics.

Advantages With the advent of the integrated-circuit piezoelectric low impedance accelerometer, the above cable and dropping problems were solved. An integrated circuit is built into the accelerometer housing, as shown in Figure 3.4, and puts out a voltage proportional to acceleration. Only a battery operated power supply is required to

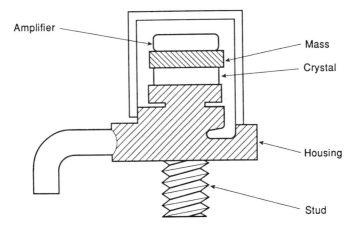

Figure 3.4A. Type Accelerometer

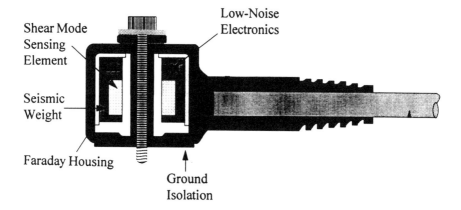

Figure 3.4B. Shear Accelerometer

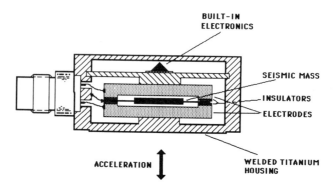

Figure 3.4C. Capacitive Accelerometer

drive the circuit. The cables can be over 200 ft long without attenuation of the signal. It is possible to purchase an accelerometer that is isolated from its surroundings to eliminate the possibility of problems with ground loops or other electronic interference as described in Chapter 4.

Many accelerometers give flat responses down to 1 or 2 Hz and up to 10 or 20 kHz. If one is interested in lower frequencies, it is possible to use a special seismic low-frequency accelerometer.

Most accelerometers are designed such that the crystal is normally in compression. If the surface on which the accelerometer was mounted were subject to small oscillations in surface temperature, a spurious signal could result. In the unlikely event that this is a problem, a design in which the crystal is in shear can be purchased. Such a design is insensitive to base strain caused by thermal variations of the mounting surface.

Disadvantages The major disadvantage of an Integrated Circuit accelerometer is that the components in the built-in amplifier are damaged at temperatures over 250°F. One is then forced to use a charge-amplifier-type accelerometer with all of the above-mentioned problems. Since charge amplifiers are now built that are smaller than a pencil, a solution to many of these problems is to mount one of these small units as close to the accelerometer as possible without touching the hot surface being measured.

Signal Integration

Even though modern accelerometers are significantly more practical and trouble free than velocity pickups, many people continue to use velocity pickups because they believe that accelerometers will not show them the correct information. The following section will attempt to dispel the myths and misconceptions that foster this erroneous belief.

For the purposed of this section, we shall assume that the class of machines of interest to the vibration analyst warrants the use of velocity as the correct amplitude parameter (examples: low-speed centrifugal pumps, paper-mill-dryer roll bearings, etc.)

How to Obtain Velocity-Based Vibration Data

It is wrong to assume that a velocity pickup must be used to obtain velocity readings. It is, in fact, a poor choice because of the shortcomings of this type of pickup. A far superior method is to use an accelerometer as the transducer and integrate the signal.

The term "integration" comes from the fundamental relationship between velocity and acceleration, rewritten here, solving for velocity as:

$$v = \int a\, dt$$

Since it is true that, for sinusoidal motion, the acceleration can be written as

$$a = -A \sin(2\pi f t)$$

Then

$$v = -\int A \sin(2\pi f t)\, dt = [A/2\pi f] \cos(2\pi f t)$$

Therefore, because the peak or rms value of the motion does not vary with time (only the instantaneous values are time dependent for nontransient vibration),

$$V = A/(2\pi f)$$

From the above equation, two things become apparent:

- If one has a frequency-by-frequency list of acceleration levels, the velocity spectra can easily be calculated.

- If the acceleration level is constant, the value of velocity will halve with each doubling of frequency. In the jargon of electronic engineers, this is equivalent to a rolloff in amplitude of 6 dB/octave. Note that this is a special case of a low pass filter as mentioned in Chaper 2.

In like manner, the equation for displacement is

$$X = -\iint \sin(2\pi f t)\, dt$$

or: $X = A/(2\pi f)^2 = [A/2\pi f]^2 \sin(2\pi f t)$

Two methods of integration exist in practice: digital and analog. The merits of these two techniques will now be discussed.

Digital Integration

Many spectrum analyzers are capable of performing a digital integration. The analyzer reads the time domain acceleration signal and generates the frequency domain spectra via an FFT calculation. The conversion to velocity data is accomplished in say, a 400-line analyzer, by use of the above equation. Each of the 400 acceleration levels is divided by each of the corresponding 400 center frequencies multiplied by 2π.

The advantage to this system is that either the acceleration spectra or the velocity spectra may be viewed at any time. The disadvantage is that the accuracy of each of the 400 calculations is a function of the bandwidth of each of the filter bins. An example of this phenomena is as follows:

> Suppose that a filter bin covers the range 87.5–112.5 Hz. The integration calculation for this band will involve a division by 100.0 Hz. If the actual frequency peak in that bin is at 90 Hz, the error in the calculation is 10%.

In the case of digital integration of an accelerometer signal to get velocity or displacement, a calibration correction must be made by either the analyzer or by hand. This correction changes the calibration from mV/g (which, when integrated to velocity yields the useless value of g-sec) to mV/in/sec/sec. With this correction, a single integration yields in/sec, and a double integration yields inches.

The correction is:

$$\mathrm{mV}/g / (386 \text{ in/sec/sec}/g) = \mathrm{mV/in/sec/sec}$$

or

$$\mathrm{mV}/g \times 0.00259 = \mathrm{mV/in/sec/sec}$$

Analog Integration

Analog integration uses the fact that the equation relating acceleration and velocity is equivalent to passing the acceleration signal through an analog low pass filter whose rolloff is 6 dB/octave. The obvious advantage to this arrangement is that the acceleration signal is converted to velocity as a continuous function of frequency. There are no bandwidth errors. The spectrum analyzer never sees an acceleration signal in either the time or frequency domain.

The analog integration is implemented either by connecting the accelerometer to a power supply with a built-in integrating filter or by purchasing an IC accelerometer with the analog filter built into the amplifier circuit. This device sometimes goes by the ill-sounding name velometer. When either of these devices are purchased, it is wise to check the frequency specifications, as it is not an easy matter to bring the corner frequency of the low pass filter below approximately 5–10 Hz. Also, one should check the dynamic range of the analog filter.

In like manner, one can obtain displacement from an accelerometer by passing the signal through a low pass 12 db/octave filter. This is referred to as double integration.

As a last note, remember that digital integration is invalid when order tracking because the center frequencies of the bins are unknown. Analog integration must be used. Some analyzers have this built-in capability.

The Problem of Dynamic Range

Most spectrum analyzers have a dynamic range of approximately 70 dB. If one were interested in a low-frequency range to observe something like a paper-machine-dryer bearing in the presence of a high-frequency signal of high amplitude such as a gear mesh, it would be possible to overload the analyzer's antialiasing filter due to the gear mesh signal low frequency, without having enough gain to see the dryer low amplitude data of interest. This is true even if an analysis range of, say, 100 Hz is chosen and the mesh signal is at 10,000 Hz. The antialiasing filter, remember, sees all frequencies coming into it regardless of the frequency range allowed to leave it. Thus, if a 75-Hz signal is in the presence of a 10-kHz signal that is greater in magnitude by 70 dB or more, the signal of interest will be lost to the internal noise level of the analyzer (see Figure 3.5).

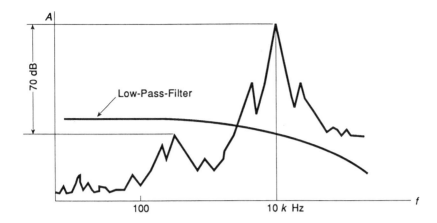

Figure 3.5. A problem of insufficient dynamic range.

The solution to this problem is to insert an analog low pass filter between the accelerometer and the spectrum analyzer. One special kind of filter that could be used is an analog integrator. In this case, the analyzer will read directly in velocity units. Of interest, the attenuation of a 6 dB/octave rolloff filter between 100 Hz and 10 kHz is 40 dB (6 dB/octave is equivalent to 20 dB/decade, where a decade is a factor of 10 in frequency). Figure 3.6 shows data taken on a gear-driven vacuum pump. Note that, with double analog integration, the pump running-speed signal can be seen in the presence of the mesh frequency.

All of the above-mentioned reasoning can be applied to a velocity signal to yield a displacement signal. An acceleration signal can be converted to a displacement signal by double integrating.

Mounting Precautions

Although the problem of how and where to mount a transducer is covered in Chapter 4, a few comments should be made in this section as well.

- Vibration readings must be taken on a surface stiff enough that it is not affected by the pressure exerted by the probe or accelerometer. A piece of sheet metal, for example, will exhibit a completely different vibration mode when it is loaded. Such a reading would not be a valid indicator of how the machine in question is vibrating.

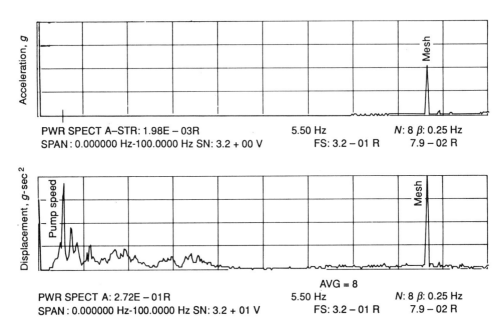

Figure 3.6. Using double analog integration to see low-level low-frequency data in the presence of high-level high-frequency data.

- All vibration readings should be taken with the transducer perpendicular to the surface of interest.

- Vibration signals containing high frequencies must be taken with the transducer tightly screwed or glued to the surface, as hand pressure cannot hold it tightly enough to the surface for it to correctly follow high frequencies.

- Magnetically mounting the transducer is preferable to a hand-held reading, but not as good as a hard mount as far as high-frequency response. The magnet will rock if put on a curved surface, reducing its usable frequency range. Some magnets must mount on a very flat, smooth surface. Others are less sensitive. A thin film of oil or grease generally increases the usable frequency range of a magnetic mounting.

- Although most magnets are usable up to 3–5 kHz, it is now possible to purchase a "super magnet" from some accelerometer manufacturers which, if mounted on a flat surface, is good to 10 kHz. Such a magnet, although quite expensive, is an invaluable time saver in the periodic monitoring of machinery having important data over approximately 3 kHz.

Force Transducers

As will be seen in the discussion of transfer functions in Chapter 6, a motion transducer is used to determine the response of a system excitation. The frequency spectrum of the excitation must also be measured. This is usually done on mechanical structures through the use of a force transducer.

A piezoelectric force transducer is similar to an accelerometer without the mass. A crystal is mounted such that any force applied to the system either compresses or stretches it. The crystal generates a charge (positive or negative according to whether it is in compression or tension) proportional to force. For the same reasons as given above for using an IC accelerometer, it is advisable to use an IC force transducer.

Remember that a force transducer only sees the force that passes through it. In a shaker test, for example, all the energy imparted to the test item must pass through the force transducer. If a force hammer is used for impact tests, the transducer must be between the hammer handle and the tip. Figure 3.7 shows an engineer using a 12 lb. PCB force hammer on the dryer section of a paper machine. The purpose of the test was to anticipate any potential problems due to a machine speedup.

The same mounting precautions as discussed above for motion transducer should be taken for the force transducer. Special attention must, therefore, be given to the selection of a force hammer for impact testing. By the design of the force transducer and hammer structure, each size hammer has a different usable frequency range. The smaller the hammer, the higher the frequency range. The author has found that a 1-lb hammer is of adequate size to excite large turbines, while still having good frequency response at a few kilohertz.

Microphones

Microphones are not only useful for simple sound measurements, but are sometimes used as machinery health indicators as a kind of implied vibration pickup in locations where it is not

Figure 3.7. Using a force hammer to search for natural frequencies in the dryer section of a paper machine.

convenient to mount an accelerometer on a machine surface. This technique assumes that the surface of the machine acts as a good speaker diaphragm over a fairly wide frequency range. Coherence, described in Chapter 6, should be used to verify this assumption before wagering the life of a machine on it.

A technique known as acoustic intensity, which employs either two microphones or a microphone and a motion pickup is sometimes used to locate noise or surface vibration where coherence will not work because the structure is so stiff that everything in the structure is coherent to everything else, as inside the block of a diesel engine. See Chapter 9.

Several different types of microphones are available to change the incoming pressure wave of a sound into a voltage that can be preamplified (in the case of electret and ceramic microphones) and fed into a spectrum analyzer.

Electret Microphones The electret microphone uses a thin plastic sheet with a conductive material coating its outer surface as a diaphragm. The inner surface of the diaphragm rests on the raised points of a perforated metal backplate. A capacitor whose characteristics vary with the pressure incident on the diaphragm is thus created. The change in capacitance causes a voltage proportional to pressure. Since the capacitor is self-polarized, the electret microphone is quite reliable over a wide range of ambient conditions.

Condenser Microphones The condenser microphone uses a diaphragm/air gap/backplate system to form a pressure-sensitive capacitor. Since a high polarizing voltage is required across the capacitor, the condenser microphone is more sensitive to field temperature and humidity conditions.

Ceramic Microphones The ceramic microphone uses a diaphragm pushing on a ceramic piezoelectric crystal and is thus quite similar to an accelerometer or force transducer. Therefore, this device is rugged, stable, and relatively insensitive to ambient conditions. The usable frequency range tends to be lower than condenser or electret microphones.

Generally, a smaller diameter microphone has a higher frequency response than a larger one. A 1-in ceramic unit, for instance may have a flat response to approximately 1,500 Hz, whereas a similar 0.5-in unit can go to approximately 4,000 Hz. For good low-frequency response, a wind screen should be used in the vicinity of machinery with windage or outdoors.

Microphones are designed for either random incidence or directional incidence of the sound wave on the microphone. The sensitivity of a given design to the incidence of sound waves should be considered before measurements are made. This information can be obtained from the manufacturer for a given model number.

Hydrophones and Pressure Transducers

Hydrophones and pressure transducers are basically waterproof piezoelectric microphones like the ceramic microphone described above, but designed for various pressures and fluids. Note that these devices are generally designed to read only the non-dc component of the pressure. This is done by providing a vent hole across the diaphragm of the transducer. Thus, if a pressure transducer was placed in a pressure line next to a pressure gauge reading 100 psi, for instance, the transducer readout would see only the higher frequency perturbations about the 100-psi constant pressure, not the 100-psi value itself.

Microphones are generally calibrated before and after each use with a sound calibrator, which fits over the head of a microphone and generates a calibrated signal at known frequencies and amplitudes. Hydrophones and pressure transducers are less sensitive to deterioration or damage and are therefore usually used by simply entering the manufacturer's sensitivity calibration value into the spectrum analyzer as is done for motion pickups.

Transducer Specifications

Before using a transducer to take measurements, certain specifications should be checked to verify that the device will output what the user expects to see. Include the following in your considerations:

Sensitivity Constant Remember that, whatever transducer is used to make the measurement, the quantity seen by the instrument is usually in volts. One must know, therefore, how many volts a given transducer puts out for a given unit of measurement. An accelerometer, for instance, is calibrated in terms of mV/g. A unit with a sensitivity of 100 mV/g will send out 100 mV (1 mV = 0.001 V) for each g of acceleration. Note that it doesn't matter if the units of sensitivity are mV rms/g rms or mV Pk/g Pk. In the case of an accelerometer, the author has found that a sensitivity of 100 mV/g always seems adequate for rotating machinery. Building vibration studies sometimes require a 1–V/g unit. Although the author has sometimes been forced to use a 10-mV/g unit due to its higher frequency range or lower mass, there have been times where a low-vibration device, such as an x-ray tube has exhibited such low acceleration levels that the output of a 10-mV/g accelerometer was too low to take advantage of the full dynamic range of the spectrum analyzer. The low-level data is masked by the internal noise level of analyzer.

Frequency Range The usable frequency range of a transducer is that range over which the sensitivity constant remains fixed, usually within 5% or 10% of the nominal value. This frequency range is limited by natural frequencies within the transducer itself. It is possible to get an accelerometer with a frequency range of 1 Hz–10 kHz, for instance, but most accelerometers only have a flat response from approximately 10 Hz to 3 or 4 kHz. Smaller units (2–10 mV/g) are usable to 20 kHz. The frequency range of impact hammers are a function of size.

For most machinery applications, a 1-lb hammer offers a good compromise between the possible size of the impact and the frequency range.

Amplitude Linearity All transducers have a range of values that they can measure. For values below or above this range, the output of the transducer is no longer linear with the amplitude of the property being measured. The upper limit of linearity of an accelerometer is 25–100 g. Usually, the lower limit is approximately 100 dB below that. A microphone is quite easily overdriven and a proximeter fails to see its target at more than a gap of 0.25 inches.

Temperature Linearity As a transducer gets too hot or too cold, its linearity falls off. Some constructions of microphones are very sensitive to this. IC accelerometers fail to operate at all over about 250° F. One must know the temperature of the environment in which the transducer will operate and choose one that is linear in that range.

Calibrating a Spectrum Analyzer for a Specific Transducer

Because electrical instrumentation uses voltage as the primary indicator of amplitude, it is often necessary to recalibrate an instrument so that the readout is in units relevant to the transducer being used. In many cases, this is as simple as determining the sensitivity of the transducer and entering it into the memory of the instrument. Thus, a pressure transducer of sensitivity 100 mV/psi would require the user to enter the value of 0.1 V/psi into the instrument's memory. The analyzer would then know that, each time it measured 1 volt, its readout should be ten "engineering units" (which the user would read as 10 psi).

As mentioned in Chapter 1, the issue of reading amplitude values in peak or peak-to-peak causes some additional confusion. Note that, in the above example, 100 mV/psi is equal to 100 mV rms/psi rms or 100 mV pk/psi pk. Thus, it one were interested in reading an output in terms of peak values on an instrument designed to read out in RMS values (as most instruments do), the instrument would have to be fooled into reading out peak values by entering

100 mV rms/psi rms × 0.707 rms/pk = 70.7 mV rms/psi pk

Many newer design spectrum analyzers have the built-in capability to do this correction itself. In such a case, the 100 mV/psi number would be entered along with the selection for an output in peak units.

If it were desired to read-out values of peak-to-peak, the correction factor would be

$$100 \text{ mV rms/psi rms} \times 0.707 \text{ rms/pk} \times 0.5 \text{ pk/pk-pk}$$

$$= 35.35 \text{ mV/psi pk-pk}$$

The above calculations become more complicated when one is using a motion transducer such as an accelerometer, and intends to integrate the signal to read-out in velocity or displacement units. As an example, let us look at an accelerometer of 100 mV/g sensitivity. Since a single integration to obtain velocity is equivalent to dividing by the frequency (in units of radians/sec = 1/sec), the readout, if uncompensated for convenient units, would be in terms of g-seconds. This is because:

$$g/(1/\text{sec}) = g\text{-sec}$$

In like manner, the result of a double integration to displacement would be in units of g-sec^2, an equally useless unit measure.

In order for the analyzer to properly display useful units when integrating, one must abandon the use of g's for acceleration in favor of its equivalent value of 386.4 in/sec/sec. Thus, if we were interested in reading out velocity in inches/sec rms or displacement in inches rms, the values of sensitivity to be entered (for a 100 mV/g accelerometer) would be

$$100 \text{ mV}/g \times 1g/386.4 \text{ in/sec/sec} = 0.259 \text{ mV/in/sec/sec}$$

Table 3.3.

Noninal Sensitivity	For velocity, in/sec			For displacement, in		
	rms	0–Pk	Pk–Pk	rms	0–Pk	Pk–Pk
10 mV/g	0.0259	0.0183	0-00915	0.0259	0.0183	0.00915
100 mV/g	0.259	0.183	0.0915	0.259	0.183	0.0915

Note that, if one wanted to read velocity in inches per second peak or displacement in inches peak-peak, the additional factors of 0.707 and 0.3535 explained above would also have to be entered into the sensitivity value to be used. See Table 3.3.

Sometimes, the use of the published sensitivity of the transducer is not acceptable as a criteria for amplitude readout. The United States Navy, for instance, requires that any sound and vibration testing for MIL-STD 740 acceptance testing use a calibrator for setting readout levels. This method makes good sense in cases that use microphones and which are sensitive to temperature and humidity, but not in the case of an IC accelerometer, which tends to be more reliable than the portable shaker used for calibration. The method for sound calibration, as an example, is as follows:

Set up the analyzer and microphone in the normal manner.

If the calibrator is known to generate a tone of 100dB at 100 Hz, set the analysis range of the analyzer to something more that 100 Hz.

Mount the calibrator to the microphone and turn on the calibrator. Take a number of summation averages.

Move the cursor on the display screen to the 100 Hz peak and set the calibration control so that the amplitude at the cursor is exactly 100 dB.

The analyzer is now calibrated.

The analyst should know the characteristics of the transducers being used. The lack of knowledge of these characteristics and operating limits can lead to some embarrassing conclusions. When purchasing a transducer, study the manufacturer's data sheet. If there is some unusual condition, such as the presence of a high degree of radio activity or explosive gasses, call the manufacturer and talk about the special situation.

Figure 3.8. PCB 303A series accelerometers.

If one has inherited a stock of old transducers, it is wise to contact the manufacturer to get the specifications for the particular serial number transducers involved. Sometimes, it is wise to get the transducers recalibrated.

Chapter 3 Questions

1. Convert the following to g's RMS:
    a. 0.1 in/sec peak at 85 Hz
    b. 0.25 in/sec peak at 200 Hz
    c. 0.003 in P–P at 125 Hz
    d. 1.25 in/sec p–p at 800 Hz

2. Convert the following to in/sec peak:
    a. 0.85 g RMS at 90 Hz
    b. 0.015 in p–p at 12 Hz
    c. 0.05 g RMS at 8 Hz
    d. .02 in P–P at 300 Hz

3. Convert the following to inches P–P:
    a. 1.1 in/sec peak at 600 Hz
    b. 0.12 g RMS at 150 Hz
    c. 0.12 in/sec peak at 500 Hz
    d. 0.075 g RMS at 5 Hz

Case Study 1: A Proximeter Case History

While in the Middle East to give a seminar, GMDC was asked to examine an ethylene compressor. The unit was showing high levels of vibration at running speed as measured by the proximeter-fed panel meter in the control room. At stake was the decision to rebuild the machine.

Transducers for Vibration Measurements 71

Solution: Acceleration readings taken at the compressor bearings did not show a significant vibration problem. Further investigation showed that the shaft, in the area under the proximeter, had been nicked with a wrench during a previous rebuild—yielding a fictitious once-per-rev reading. The unit did not have to be shut down or rebuilt.

Chapter 3 Case Study 2:

Problem: A close-coupled liquid ring compressor (compressor parts built directly onto a special overhung motor shaft) ran at 1800 RPM and had 20 blades, causing an excessive audible noise at the blade frequency of 1200 Hz. The compressor supplied instrument air to a commercial ship, and was located next to the captain's cabin. Every time the compressor started during the night, the noise would wake the captain.

Test Results: The author determined that the noise at the blade frequency was exacerbated by the fact that the first natural frequency of the motor shaft-rotating elements was approximately 1200 Hz—the blade frequency of the pump.

Solution: Since the shaft/bearing frequencies of the unit were a function of the motor design, the motor was sent to the manufacturer's local repair shop. The author explained the difference

Figure 3.9A. PCB accelerometers.

Figure 3.9B.

Figure 3.10. PCB 480A10 double integrating (analog) power supply.

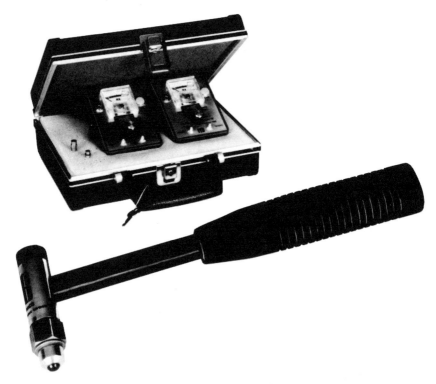

Figure 3.11. PCB GK291B01 small force hammer kit.

Transducers for Vibration Measurements 73

Figure 3.12. PCB 208B01 force transducer.

Figure 3.13. PCB mechanical impedance head.

Figure 3.14. PCB 118A05 pressure transducer.

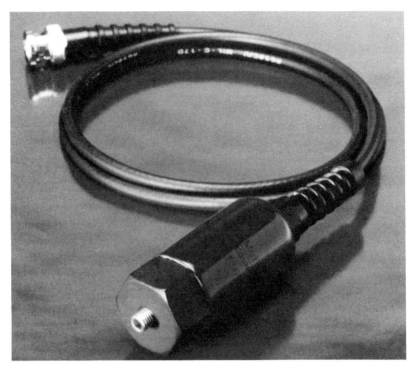

Figure 3.15. Vibrametrics 1136TB low cost industrial monitoring accelerometer.

Figure 3.16. Vibrametrics 8002ST heavy duty, chemical resistant accelerometer.

Transducers for Vibration Measurements 75

Figure 3.17. Vibrametrics 7000 triaxial accelerometer.

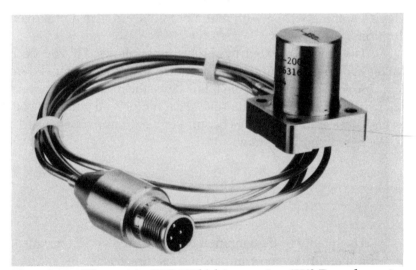

Figure 3.18. Vibrametrics VTC 100 high temperature (550° F) accelerometer.

Figure. 3.19. Vibrametrics 9002A subminiature accelerometer.

between displacement, velocity, and acceleration, and requested that acceleration motor vibrations be taken to demonstrate the natural frequency at 1200 Hz. About a week later, the answer came back: "There's nothing wrong with that motor, it's less that a mill overall vibration."

As the manager of the repair shop could not understand the reason for measuring acceleration, he helpfully offered to send the motor to the factory for evaluation. Again, the author requested acceleration readings, and again the answer came back "There's nothing wrong with that motor, it's less than a mill overall vibration."

Next, the motor was sent to the personal experimental lab of the Vice President of Engineering. The author carefully explained about displacement, velocity, and acceleration, and the V.P., astounded that his in-house people couldn't understand the concept, offered to follow up himself. Again, the answer came back: "There's nothing wrong with that motor, it's less than a mill overall vibration."

The author restated the problem as follows: "If you do not give me acceleration readings on the motor, you will be removed from our approved suppliers list." In a few days, the results were reported back by the Vice President himself: "That motor's no good for you, it has high acceleration readings because of a natural frequency at 1200 Hz. We're redesigning it now."

Summary

The quality of the vibration data gathered and, consequently, the reputation of the spectrum analyst, is directly dependent on the proper selection and mounting of the transducer. It is, therefore, necessary to use proper deliberation in making decisions regarding the transducer choice.

Proximeters are low-frequency (to dc) devices that are good for looking at shaft motion. Accelerometers have a flat response over a very wide range of frequencies. They can be integrated to yield velocity and double integrated to yield displacement (with certain dynamic-range precautions). The author takes a jaundiced view of the abundant use of velocity pickups on low-speed machinery and recommends that their use be discontinued for general machinery work.

The construction of force transducers, microphones, hydrophones, and pressure transducers have been discussed briefly. It is expected that the user of one of these devices will have the good sense to read the instruction manual that comes with them.

In order that the analyst have some feel for the size and shape of some of the transducers discussed in this chapter, we include photographs of some transducers manufactured by PCB Corporation of Depew, New York and Vibrametrics of Hamden, Connecticut (see Figures 3.8 through 3.17).

CHAPTER FOUR

Elementary Problem Diagnosis

Every moving component of a rotating machine generates a vibration signal that is uniquely its own. For example, bearings cause a set of frequencies that are different from the frequencies caused by couplings, gears, impellers, or an out-of-balance component.

By gathering the total vibration signal of the machine in question at the appropriate locations, and breaking this information down into its frequency content, it is possible to learn the effects of each mechanical component. This chapter will begin by discussing the proper locations to gather data, some simple methods to diagnose certain kinds of problems, and a detailed section on using a single-channel FFT spectrum analyzer to do problem diagnostics.

It is important to remember that any rotating machine has many of the same vibration characteristics as any other rotating machine. This is true whether the machine is a 180-rpm water pump or a 10,000-rpm gas turbine. This is why vibration analysis has been as successful as it has been. If a new set of physical laws had to be written for each new machine, the technique would be next to useless. Throughout the following chapters, keep in mind that what is said concerning the behavior of a particular pump component, for instance, is just as true for the analogous component on the high-speed compressor that is presently shaking your entire plant.

Where to Measure Vibration

For rotating machines, vibration readings are generally taken at each bearing housing in the system in the horizontal, vertical, and axial directions. This information is useful to the manufacturer of the machinery in question because the vibration of the rotating element is

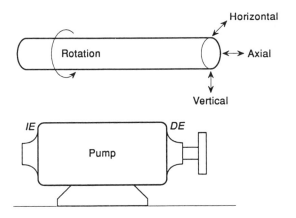

Figure 4.1. Common measurement locations.

transmitted through the bearings; this effect is more pronounced with rolling element bearings than sleeve bearings (see Figure 4.1). Further vibration readings may be taken at any location that seems relevant to the particular problem at hand.

Vibration readings should be taken with the transducer perpendicular to the surface of interest. Vibration signals containing high frequencies must be taken with an accelerometer tightly screwed or glued to the surface, since hand pressure cannot hold it tightly enough to the surface for it to correctly follow high-frequency motion. Magnetic mounting using an accelerometer is preferable to a handheld reading, but is not as good as a hard mount for high-frequency response.

Trouble-Shooting Guide

The following guide is intended to help the vibration analyst begin the attack on a problem using simple equipment and a logical problem-solving technique. If these efforts yield unsatisfactory results, then it is time to do a detailed spectrum analysis. The clever person does not make anything more complicated than it is.

Background Readings

Before assuming that a machine is vibrating excessively, it is often necessary to take readings in certain positions with and without the suspect equipment running. The vibration could be caused by some other piece of machinery transmitting vibration through piping or flooring. Good places to take data are at the bearings, at the associated piping, and on the soleplate of the suspect machine. Take readings with the suspect machine operating and compare them to readings with the machine off when the rest of the plant is operating.

If the suspect machine's operation does not significantly affect the unfiltered vibration levels, use an accelerometer and magnet to "feel" along piping and flooring until the real culprit is found or, if a 2-channel spectrum analyzer is available, do a coherence check as described in the next chapter to locate the source of any particular frequency component of concern.

Another way to determine if the suspect machine is the actual culprit is to test the machine with the rest of the plant off. These tests are typically run at night or on weekends.

Piping

Keep in mind that the mass of the machine of interest is often small compared to the mass of the piping connected to it. The problem may well be caused by the piping, not by the machine. Typically, the problems associated with piping are either due to misalignment or resonance. A simple test can be run if the piping can be disconnected while the machine is running. If the unfiltered vibration level changes significantly—either greater or lower—one of the following problems can exist:

- The piping is improperly supported and is causing either a coupling misalignment or a shift in the natural frequency of the driver–driven machine–base–piping system such that it is being excited by some forcing mechanism in the system. A possible solution is to use flexible pipe connections to change the stiffness of the system.

- The piping is causing internal misalignment of the machine (such as bearing bracket deflections). Flexible pipe couplings must be used in this case. Beware of the effect of thermal expansion in the piping.

- Some other machine, which is connected to the pipe, is shaking the machine in question. Take background readings or try synchronous time averaging with your spectrum analyzer.

- Some flow condition, such as water hammer, vortex shedding, or cavitation, is exciting the machine. Move the accelerometer along the pipe to determine the point of greatest excitation. Look for orifices or partially closed valves. If the flow problem is not obvious, as it often is, check the natural frequency of the pipe using 2-channel analysis to see if a resonance is being excited. Check with the Process Engineer to determine what can be done about a poor flow condition.

Bases and Support

Inadequate bases and the support of bases are frequently the cause of vibration problems in machinery. The following are typical problems and methods for checking whether these problems exist:

Poor Mounting Compare readings taken on the machine at the bearings to readings in the same direction at the machine mounting feet and on the base near the feet. If the levels at the mounting feet are significantly greater than on the base near the feet, the machine is not being properly held to the base. Torque the hold-down bolts and recheck the vibration.

Poor Base If the vibration readings on the base near the machine feet are significantly higher than the readings elsewhere on the base, the base is not sufficiently rigid. This is a common occurrence in bases fabricated of wide flange beams because the resistance to bending about the vertical axis is only about 10% of the resistance to bending about the horizontal axis.

One can add sides to the flanges of the beams to make them into rectangular cross-section beams, replace them with rectangular cross-section beams, or weld diagonals or X bracing to the base to increase its torsional stiffness. All welds should be continuous for strength and all bracing should tie into the web and flanges of the wide flange beams as shown in Figure 4.2. Do not fill the base with grout. Although this seems to solve a weak base problem, high-frequency vibrations will powder the grout-steel interface in approximately 3 years. The original base weakness will reappear but, since the base is now filled with grout, a welded repair will not be feasible.

Another sign of a poor base is the appearance of high-level coupling misalignment signals as described later in this chapter. Attempts to realign the system will show temporary results, at best.

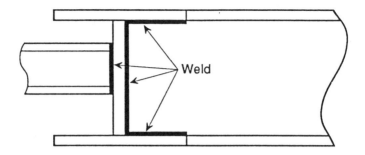

Figure 4.2. Correct (right) and incorrect (left) method of bracing wide flange beams.

Poor Base Support If vibration readings on the base where it is mounted to the floor is more than 1.5–2 times the unfiltered vibration in the concrete or steel floor near the hold-down bolts, the base is not properly shimmed, grouted, or mounted to the floor. Many bases require good floor support for rigidity. These bases are called "non-self-supporting."

Sometimes, excessive vibration is caused by poor base support. To test for this, take vibration readings on the machine of interest while someone tightens or loosens the base hold-down bolts. Before running a test of this type, evaluate the safety of doing so.

Warped Base A weak base, due to improper shimming and grouting, can sometimes take on a permanent set, which has the effect of warping the machinery mounting surface. In such a case, alignment will be impossible to maintain. It has been observed that the vibration levels taken at the bearings of a machine actually can drop as the base hold-down bolts are loosened. This is because the base has been allowed to straighten out. If reshimming and regrouting the base does not relieve the problem, a new self-supporting base may be required.

Electric Interference Electrically caused problems tend to disappear instantly as soon as the current supply is removed. Mechanical vibration problems drop slowly in level and frequency as the machine gradually rolls to a stop. It is therefore possible to determine an electric problem by watching the 60- and 120-Hz component of the spectrum as the current to the machine is turned off. The level will instantly drop if the problem is electric. The electric interference could be a small ground loop that would not affect an isolated accelerometer or a potential of several hundred volts across the bearings of a machine (causing etching of the races). A simple voltmeter may be used to check for this problem.

Single-Channel Spectrum Analysis

A typical machine train is made up of an electric motor or turbine, one or more flexible couplings, and a rotating driven machine designed to accomplish some task, such as a pump to move liquid. Occasionally, a gearbox, fluid drive, or belt drive is employed to operate the pump at a speed that is different from the driver speed. All of these pieces of machinery exhibit certain vibration characteristics, which are caused by various mechanical components in the machinery. The remainder of this chapter will show how to look at a vibration spectrum (plot of amplitude versus frequency) obtained on an FFT spectrum analyzer and determine which component of what machine is causing an operational problem in the system. To do this, it is necessary to discuss the notion of forcing frequencies.

Forcing Frequencies

An operating piece of rotating machinery can exhibit several different kinds of events in the time it takes for the shaft to make one revolution. These events have the property that they *must* occur in each revolution of the shaft by the very nature of the design of the machine and the task the machine is meant to perform or the manufacturing procedure with which it is made. One can calculate these frequencies and look for them in the vibration signature of the machine. Once these frequencies are found, the question of whether the levels attached to these frequencies are too high or normal must be addressed. One must first locate the forcing frequencies.

The following are some of the frequencies that result from components found in a great many rotating machines. Other frequencies may well exist in a particular piece of equipment. If the reader tries to understand the reasons for the existence of each of the following, he will quickly gain competence in the ability to calculate forcing frequencies generated by any mechanical mechanism for which he has a good physical understanding.

Rotational Speed

A signal at the rotational speed (shaft rpm) of every rotating mass in the system will be found to exist whenever the machine is running. A common cause of a peak at rotational frequency is unbalance. The unbalanced force generated by machining errors in a rotating mass is proportional to the shift in the center of gravity from the geometric center of the shaft, the amount of unbalance, and the square of the speed (see Figure 4.3). Thus, a machine may exhibit acceptable vibration levels at one speed but severe levels at a higher speed. Note that the unbalanced force vector points radially out from the center of mass and rotates synchronously with the mass. If a vibration pickup were mounted on a bearing of a machine in, say, the vertical direction, the output of the pickup would peak at the instant that the rotating unbalance vector was pointing directly at the pickup. If the phase difference between this peak and the occurrence of another synchronous event, such as a keyway passing a keyphasor

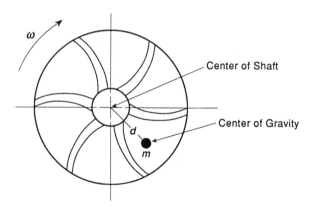

Figure 4.3. Unbalanced force = $m \times d \times w^2$.

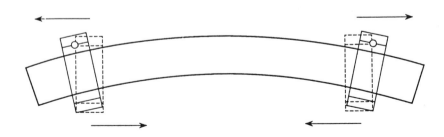

Figure 4.4. A bent shaft.

or a strobe light firing to illuminate the key in a given position, is measured, it becomes an easy matter to deduce (with the help of trial weights) the location of the heavy spot and its magnitude. The addition, then, of the proper mass 180° from the heavy spot will properly balance the rotating mass. Although it is theoretically possible to completely balance a rotating mass for any given speed, the sad fact is that nothing man does can be done with infinite accuracy. We simply balance a rotor such that the resultant vibration level is not harmful to the machine and continue on our way.

Other common causes of a 1 × shaft speed signal are misalignment of couplings, bowed shafts, and eccentric or unbalanced belt or gear drives. It is sometimes possible to differentiate among possible causes of 1 × rpm by observing phase differences across the machine or noting the direction of maximum vibration level with respect to the geometry of the system. A bowed shaft, for instance, causes the bearing housings of a machine to move 180° out of phase to each other in the axial direction as they are forced together and apart by the bent shaft (see Figure 4.4).

Uneven loading, as from eccentric belts or gears will be a maximum in the direction of pull. The vector will not rotate. The level of vibration will vary with load, not necessarily with speed.

Flexible Couplings

Flexible couplings cause several different possible forcing frequencies. As mentioned above, a misalignment may show up as 1 × shaft speed. One would expect this, because misalignment also can cause high coupling unbalance forces. The classic indicator of coupling unbalance is a 2 × rpm frequency, frequently highest in the axial direction. Coupling misalignment exhibits an approximately 180° phase shift across the coupling.

Another indicator of a coupling problem is exhibited in several different coupling designs. These indicators are low-energy spikes at frequencies equal to the number of coupling pins, teeth, or grids multiplied by the running speed. These frequencies give a very early warning of incipient coupling failure and should not be ignored. Note that since these levels are quite low, they may be missed, unless the spectra is examined using a logarithmic amplitude scale. The earliest warning of an incipient coupling failure ever obtained by the author was the existence of this forcing frequency 7 months before the coupling failed and northern Arizona lost power for 6 hours.

A loose coupling is likely to cause sidebands about frequencies, such as the blade frequencies and mesh frequencies of the

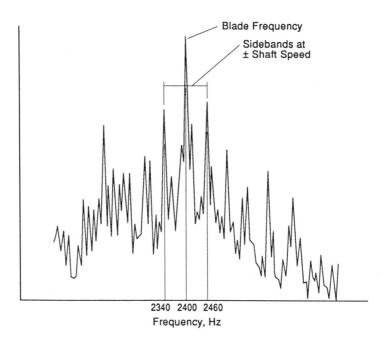

Figure 4.5. Sidebands about a blade frequency.

system. This is caused by the fact that a loose coupling does not drive the rotating equipment at a uniform speed, but, rather pulses at multiples of the shaft speed. Therefore, a signal with a well-defined blade frequency and equally spaced sidebands of +/− N × shaft speed typically means that the coupling is loose (either from a poor fit on the shaft or from worn coupling components). N is an integer multiple such as 1, 2, 3, etc. See Figure 4.5.

Blade Frequencies

Every time a rotating blade moving a fluid passes near a discontinuity, such as the tongue of a volute, one expects an impact. It therefore follows that a 6-bladed impeller, for example, would have a blade frequency equal to 6 times the shaft speed of the impeller. Multiples of the blade frequency would also be expected to exist because there are multiple discontinuities in the surface of the casing and the impacts, not being purely sinusoidal, will cause a wave shape made up of many terms of a Fourier series. One should expect to see the blade frequency in an operating pump, compressor, fan, or turbine. An extremely high-amplitude level at this frequency, however, could possibly mean, in the case of a pump, for instance, that either there is insufficient clearance between the blade tips and the tongue of the volute or that the pump is stalled. For any rotating bladed machine, an abnormally high amplitude at the blade frequency is indicative of a problem in the hydrodynamic flow path about the blades of the machine.

Gears

Gears generate a mesh frequency equal to the number of teeth on the gear multiplied by the rotational speed of the shaft driving it. Notice that the mesh frequency of a bull gear times its shaft speed is equal to the mesh frequency of the pinion gear times its shaft speed. See Figure 4.6, where gear mesh = 100 × (1800/60) = 300 × (600/60) = 3 kHz.

If one were to take two gears of 1.00 : 1.00 ratio, mesh them together, and rotate one of them by 360°, the other gear might rotate 359.99° or 360.01°. The difference in angular travel is called tooth error, and results from indexing errors when the gears were hobbed. One of the major differences between a set of quality gears and a set of relatively inexpensive gears are the tolerances to which the gears are machined.

A high vibration level at the mesh frequency is typically caused by tooth error due to poor machining, wear of the meshing surfaces, improper backlash, or any other problem that would cause the profiles of meshing teeth to deviate from their ideal geometry.

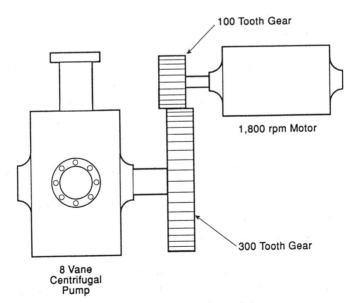

Figure 4.6. A typical gear drive.

Thus, it is possible for unloaded gears to fail to show any mesh frequency at all because the tooth errors are covered up by conforming changes in the oil film. When the gears begin to be loaded, the oil film can no longer compensate for errors and the classic mesh frequency pattern appears.

Sidebands of the mesh frequency occur for the same reason as any other sidebands—a modulating rotational motion. in the case of a gear drive, this modulation results from a failure of mating teeth to impact one another at the proper time. Suppose, for example, one had two meshing gears of 100 teeth each turning at 1 Hz. Each pair of teeth should meet exactly 0.01 sec after the previous two teeth have met. If the spacing between adjacent teeth on one of the gears is too large, the mating tooth on the other gear would have to wait too long for the impact, causing that gear to slow down. The spacing on the next pair of teeth, in order to maintain an average impact rate of one every 0.01 sec, will occur sooner than normal, causing an acceleration of the mating gear.

Gears generate a large number of possible sidebands about the mesh frequency. Sidebands at the mesh frequency plus or minus multiples of the pinion speed and sidebands of the mesh frequency plus or minus multiples of the bull gear speed are typically due to such causes as eccentric gears or nonparallel shafts, which allow rotation of one gear to "modulate" the speed of the other gear. Since the motion that occurs is a combination of both amplitude modulation and frequency modulation, the resulting pattern of sidebands about a mesh frequency is typically a set of asymmetrical peaks at multiples of the running speed of both of the gears.

It is frequently necessary to differentiate between high vibration amplitudes at mesh frequencies and a high-energy content in the sidebands as they indicate two different possible problems in the gearing. A high level at the mesh frequency can indicate an interference fit between mating gears, for instance a tooth error-type problem, whereas a large number of high-level sidebands can indicate nonparallel shafts (due to excessive shaft deflection, misaligned shafts, gear tooth, or housing cracks), an eccentric gear, or some other type of modulating motion.

It is advisable to look at gear signals using the frequency expansion (zoom) capability of a spectrum analyzer. The following gear example shows the necessity of expansion:

> If one has a gear set with a 40-tooth gear running at 60 Hz driving a 300-tooth gear at 8 Hz, the following can be seen:

- Mesh Frequency = 40 × 60 = 300 × 8 = 2,400 Hz
- Sidebands at

$$+/-N \times 60 \text{ Hz} = +/- 60, +/- 120, +/- 180, \text{ etc.}$$

and at

$$+/- N \times 8 \text{ Hz} = +/- 8, +/- 16, \text{ etc.}$$

To see the mesh frequency, the analyzer must be in the 5 kHz range. The bandwidth of a 400-line analyzer in this range is

$$\beta = 5,000/400 = 12.5 \text{ Hz}$$

One cannot differentiate sidebands spaced at 8 Hz intervals with a set of filters 12.5 Hz wide. A frequency span of 500 Hz is required. The expanded bandwidth will be

$$\beta = 500/400 = 1.25 \text{ Hz}$$

Thus, use of the frequency expansion option has permitted the detection of the bull gear sidebands, and an accurate determination of a gear problem is possible.

There are still more forcing frequencies that can be caused by gears. One is the *hunting tooth* frequency, in which a tooth with a large geometric flaw on one gear impacts on a tooth with a large flaw on the other gear on a repetitive basis. if the gears were a 1.00 : 1.00 ratio, the hunting tooth frequency would coincide with the running speed. If, however, the gears had a 1.00 : 1.01 ratio, the bad teeth would impact only each 101 revolutions of the slower shaft. If you hear an occasional thump every few minutes on large, low-speed gears, it is probably a hunting tooth frequency.

Gears can also generate *ghost frequencies*, which are caused by the method of manufacturing the gears. Suppose that a gear blank is held in a three-jaw chuck for machining and the operator is overzealous in tightening the chuck. The gear teeth will be cut at a uniform diameter but, when removed from the chuck, will stress relieve at three symmetrical places to form a "triangular gear." Expect to see a large component at 3 × running speed. Other imperfections in the method of machining a gear, such as the number of cutting blades in the hobbing process, can generate other ghosts.

Table 4.1. Diagnosing Gear Problems

Symptom	Problem
High Amplitudes at the Mesh Frequency	Tooth Error Wear Improper Backlash Overloaded Gears
High Energy Levels at Sidebands	Eccentric Gear Non-parallel Shafts Gear Tooth Cracks Gearhousing Cracks Other Modulations (such as worn couplings)

Rolling-Element Bearings

Rolling-element bearings cause several frequencies, which can be calculated. The equations below indicate the frequencies one would see due to a race defect, a ball defect, and an unbalanced cage:

inner race fault: $1/2\ zf\ (1 + (d/e) \cos a)$

outer race fault: $1/2\ zf\ (1 - (d/e) \cos a)$

ball fault: $f(e/d)\ (1 - (d/e)^2 \cos^2 a)$

cage unbalance: $1/2\ f\ (1 - (d/e) \cos a)$

where f is the rotational speed in Hertz, d is the ball diameter in inches, e is the pitch diameter in inches, a is the contact angle in degrees, and z is the number of balls (see Figure 4.7). These frequencies may not always be visible because a bearing tends to be a low-energy device compared to the remainder of the rotating machine (up until the time that the bearings are well on their way to destruction), so that the above noted equations may or may not be helpful.

The chances of picking up an inner race fault are small unless the load direction of the bearing coincides with the location of the accelerometer. Any impact between the inner race and a rolling element must transmit across the rolling element, through an oil film, through the outer race, and through the bearing housing, to the accelerometer. The difficulty of the transmission path from the inner race to the outside world helps to explain why an outer race fault tends to be the easiest one to observe.

A spall on a ball of a ball bearing will not always contact the inner and outer race, because sometimes the spall may be oriented in the axial direction and miss a race. Therefore, this effect may be difficult to pick up.

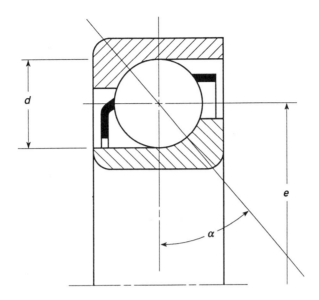

Figure 4.7. The geometry of a bearing.

Some papers on the vibration monitoring of bearings state that bearings are least likely to fail because of ball problems. It is far more likely, however, that ball failures are least likely to be observed until severe damage has occured on the races as well as on the balls. This problem of observation does not appear in roller bearings because the rollers always maintain the same orientation with respect to the races. By the time the bearing cage is deteriorated enough to cause a measurable amount of unbalance at the accelerometer, it is likely that total failure is only a short time away.

Note that a bearing that is loose in its housing will spin at about half shaft speed, causing a peak at that frequency. Since, by this time, the internal structure of the machine is no longer properly aligned, a 2 × rotation peak may also be seen.

A further confusing phenomena in bearing analysis that the above equations fail to allow for is when bearing components slip along each other on an oil film, rather than maintaining rolling contact. Although slippage is usually only a few percent, it has been observed to be almost fifty percent during some modes of operation of a gas turbine engine in flight.

There are other techniques for determining the condition of a bearing. One method is to look at the natural frequency of the accelerometer being used. An imperfection in the bearing, being of the nature of a short duration pulse (approaching an impulse) appears as white noise in the frequency domain. Since white noise contains all frequencies, it also contains the frequency that equals the natural frequency of the accelerometer. Thus, the imperfection in the bearing will excite this natural frequency. Because of the amplification of the white noise at the accelerometer resonance, the impact will act as an early indicator of bearing failure. Accelerometers tend to have resonances in the 30–90 kHz range. Therefore, it is important to have a spectrum analyzer with a 100 kHz range for this shock-pulse technique. Specific devices to measure only this one phenomena exist. A more detailed discussion of the phenomena can be found in Appendix B, Pulse Theory.

If a frequency expansion capability is built into the spectrum analyzer, it is possible to look at the natural frequency of the accelerometer with sufficient resolution to see sidebands at the various bearing frequencies calculated above. It should be noted that this method only works if the accelerometer is mounted to the bearing in such a rigid manner as to transmit high-frequency impulses (the accelerometer should be screwed directly to the bearing housing). Also, be aware that some accelerometers, in order to extend operating range, use a low pass filter in the IC circuit to reduce the effects of the resonance. These accelerometers cannot be used for shock-pulse readings.

Another variation of the above method is to determine the area under the spike-energy curve. The value obtained from this measurement is proportional to the area of the surface damage rather than the depth, as in the spectrum analysis or shock-pulse methods. After the initial stages of failure, damage area increases more than damage depth.

One major problem with any technique relying on the generation of white noise to locate bearing problems is that there may be other sources of white noise in the system. A classic example is a steam leak. Anyone who has tried to check bearing condition on dryer bearings in a paper mill knows that all of the bearings appear to be terrible. This is because all of the bearings tend to be near steam leaks from the dryers. The shock-pulse meter cannot differentiate between white noise from a steam leak and white noise from a bearing fault. The only method that has any chance at all in this application is spectrum analysis using the calculated forcing frequency equations.

Another method of monitoring bearings, which is just beginning to be tried, is to permanently mount a proximeter with an upper frequency limit of approximately 10 kHz such that it detects slight motions of the outer race of the bearing. It is said that the impact of a fault is adequate to set this race in motion.

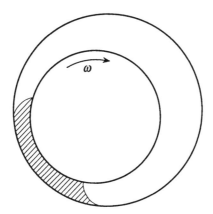

Figure 4.8. A sleeve-bearing oil wedge.

Fluid-Film Bearings

The center of a shaft does not remain at the geometric center of a fluid-film bearing during operation due to relatively large bearing clearances and the thickness of the oil-film wedge supporting the shaft. It is therefore possible for the shaft to whirl inside the bearing when various instabilities in the oil film develop. These whirl frequencies range from approximately $1/3-1/2 \times$ shaft speed and indicate oil-film stability problems (see Figure 4.8).

The instability can also be seen as a figure-eight pattern on a shaft-orbit plot, which may be caused by excessive clearances from bearing wear, improper oil viscosity due to improper pressure, temperature, or lubricant condition in the wedge, or improper bearing surface area. If the problem is caused by a design flaw, rather than wear, there are several solutions: (1) heavily load the oil wedge by using tighter clearances, (2) increase the load/unit surface area of the bearing surface, (3) use tilting pads, or (4) axially groove the bearing to break up the laminar flow characteristics of the wedge.

When monitoring oil-film bearings, it is wise to periodically check the condition of the oil by sending samples of the oil for ferrographic checks on a regular basis. Remember that a rolling-element bearing tends to have a relatively gradual deterioration, whereas oil-film bearings often fail almost instantaneously. Continuous monitoring with proximeters and temperature transducers are not amiss on a critical machine.

Cavitation

Cavitation is caused by the implosion of vapor bubbles that were formed in a fluid-flow region where the local pressure of the liquid dropped below its vapor pressure. The implosions take place when the fluid pressure has increased above the pressure level that can sustain the existence of the bubbles. These implosions tear out tiny pieces of the metallic surface near which they implode. Figure 4.9 shows a schematic representation of the cavitation process.

Since the bubble collapse is a high-energy, short-duration event, it can be detected in the same manner as a bearing fault (using the resonant frequency of an accelerometer) or by noting a sudden increase of level across the entire frequency range of interest. This increase will be of a low-energy content, such that the peaks of a spectra will probably remain unaltered but the valleys will tend to fill in. A more detailed discussion of the appearance of cavitation will be made in Appendix B in the section on Pulse Theory.

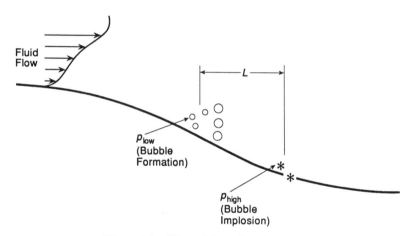

Figure 4.9. *The origins of cavitation.*

Looseness and Rubs

The motion of a bolt loose in its hole, rather than exhibiting a smooth sinusoidal motion of rocking back and forth, will be truncated by the sides of the hole. Therefore, the motion will be something like a square or trapezoidal wave. The signal will have many harmonics of the excitation frequency (usually shaft speed). Impacts of the bolt against the walls of the hole may also cause the addition of white noise to the spectra and kick off any natural frequency in the system.

Rubs can be considered a dry-friction whirl. The spectra caused by a rub may have peaks at frequencies of

$$f = 1/n$$

where $n = 2, 3, 4, \ldots$. A shaft orbit plot will appear as a series of loops joined end to end to form a circle. The direction of the orbit will be opposite to the direction of shaft rotation.

Note that a shaft rub may cause enough localized heating to bow the shaft, evoking the signals described above for unbalance and misalignment. A rub can also generate enough white noise to set off natural frequencies.

Motors

Motors use electromagnetic forces, as well as mechanical forces, and, therefore, exhibit some characteristics that differ from hydrodynamic or purely mechanical rotating machines. For example, a motor shaft may bow due to excessive localized heat buildup from shorted laminations. The appearance of this phenomenon will be the same as an unbalance as mentioned above under rotational speed, but balancing obviously is not the correct solution to the problem.

Also keep in mind that, unlike a purely mechanical machine, which can be modeled as a set of springs and masses for natural frequency studies, a motor also has electromagnetic springs, which are a function of loads, gaps, and magnetic centers.

A defect in a motor stator can be seen as a peak at twice the current frequency. This is usually 120 Hz in the United States. If one watches the spectrum on an FFT analyzer in the instantaneous or exponential averaging mode (with as few averages as possible), it will be seen that an electrical problem at 120 Hz will disappear when the motor is deenergized, whereas a mechanical problem will change in amplitude and frequency as the rotating mass rolls to a stop.

A rotating electrical defect such as a broken bar will show up as sidebands of

$$(\text{rpm} +/- 2 \times \text{slip frequency})/60$$

The slip frequency is the difference between the synchronous speed of the motor and the running speed. In the United States, the possible synchronous speeds are given by

$$120/\text{number of poles in the motor, in Hz}$$

For normal properly loaded induction motors, the slip frequency is usually approximately 5% of the running speed.

Summary

No attempt has been made here to describe all of the possible forcing frequencies in the world. The hope is that the reader will grasp the kinds of mechanisms that cause the above-mentioned signals and be able to add to this catalog, applying similar reasoning to specialized machinery.

Warnings concerning natural frequencies have been made throughout this chapter. One should always ask whether a level at a particular forcing frequency is high because of a problem with the forcing mechanism, or whether the forcing mechanism is exciting a natural frequency. The next chapter will instruct the reader on the methods of answering this important question.

Case History: An HVAC Warranty Problem

A Fortune 500 drug manufacturer was building a new research facility. Many of the experiments performed in such a facility run for years and can be ruined by temperature variations as small as four degrees. Temperature control in such a building is critical. Because of delays in the construction of the building, dozens of air conditioning units and fans were stored outside for months awaiting installation. Upon installation, the manufacturer inspected the equipment, changed a few bearings and belts, and declared everything to be fit. GMDC was called in to verify this before the expiration of the equipment warranty.

Result: Upon doing a detailed spectrum analysis on the vibration characteristics of the various kinds of equipment, GMDC located a myriad of failing bearings, bad belts, misalignments, and motor control problems which had to be addressed before the building could be occupied.

Case Study: A Coupling Problem

The Problem: A power plant in the Southwest used an integral gear driven liquid ring vacuum pump for condenser exhaust service. The unit was driven via an 1800 RPM induction motor and a grid-type coupling to drive the integral gear set and 16 bladed vacuum pump. Plant personnel complained about the noise of the unit and asked the author to investigate.

The Measurements: Vibration readings showed reasonable amplitudes at harmonics of the running speeds of the motor and pump, normal amplitudes at harmonics of the blade frequency (16 × pump speed, etc.), high amplitudes at harmonics of gear mesh and sidebands, and a small peak at 55x motor speed. In addition, each major peak showed sidebands at +/− motor speed.

The Investigation: A comparison with similar pumps manufactured by the same company showed the gear mesh peaks to be of a normal amplitude for the overhung design of the gear shafts. It took extensive investigation of the paperwork for the construction of the pump package to learn that the grid coupling had 55 'bumps' on the coupling halves to match the steel grid. Further, it was reasoned, the motor speed sidebands seen about each peak could be caused by looseness in the coupling.

A report was written stating that:

a. The gears were in good condition, and that the audible noise and vibration from the gears was normal.

b. The 55x motor speed peak and the motor speed sidebands were an indication of probable coupling wear.

The Results: The power plant, confident that the gears were not failing, ignored the second finding of the report. Eight months later, the coupling failed, and the northern part of the state lost power for six hours.

CHAPTER FIVE
Dual-Channel Spectrum Analysis

In the previous chapter, spectrum analysis was shown to be a powerful tool in determining the condition of various components inside a given machine. In Chapter 6, we will show the advantage of monitoring these levels on a regular periodic basis. This chapter discusses techniques that will greatly enhance the diagnostic capabilities of spectrum analysis, but do not lend themselves to periodic monitoring.

It has long been known that the measurement of the vibration levels generated by rotating machinery, such as pumps, gears, turbines, and so on, is a cost-effective endeavor. Maintenance personnel are beginning to learn that spectrum analysis for problem diagnosis enhances these benefits greatly. Less well known are the benefits of using a dual-channel spectrum analyzer for problem diagnosis in the plant. This is, perhaps, due to the size, weight, and cost of early dual-channel analyzers. Improvements in electronic technology have permitted design enhancements in current analyzers to the point where dual-channel units are about the same size, weight, and cost as single-channel analyzers. The time is right, therefore, for plant maintenance and reliability personnel to learn the attributes of dual-channel analysis.

To discuss the advantages of dual-channel analysis, it is necessary to understand natural frequencies, transfer functions, coherence, and coherent output power. Although some mathematics will be used, the explanations of these properties will rest heavily on physical interpretations.

Natural Frequencies

All forcing frequencies, including those described in the previous chapter, share the characteristic that they are self-generated. When you turn a machine on, the forcing frequency appears; when you turn the machine off, the forcing frequency goes away. If, in fact, the machine changes speed, the forcing frequencies change proportionally, requiring order tracking. Forcing frequencies often have the property that they can be rather easily calculated if the physics of the particular forcing mechanism is properly understood. They result from the purposeful design and the imperfect manufacture of machinery.

Natural frequencies (or resonances) are quite different in character. They are due to the nature of the structure of the machinery, including the piping and the support system. They are not self-excited but can be viewed as lurking within the structure of the system, ready to cause violent reactions when excited. They result from the values of the mass, stiffness, and damping of a structure and are not a function of the operation of the machine, except in instances where stiffness is speed dependent or some electromagnetic spring exists (as may be the case in an electric motor).

Since a natural frequency, if excited, causes large increases in the amplitude of vibration at that frequency, the analyst must ask this question each time a forcing frequency with a high amplitude of vibration is identified:

> Is the amplitude of vibration high because there is something wrong with the forcing mechanism involved, or is the forcing mechanism as it should be, but exciting a natural frequency at that point in the spectrum?

This is the most important single item the vibration analyst should derive from this chapter.

When analyzing the vibration of troublesome machinery, therefore, it is often quite important to be able to determine the natural frequencies of the machine–support–piping system. Such a determination is necessary to insure that there are no forcing frequencies close to natural frequencies.

Suppose that a single-channel vibration spectra was obtained on a machine running at 1,800 rpm (30 Hz). Suppose, further, that the only thing of interest in the spectra was a high-vibration amplitude at running speed. The novice would assume that the 30-Hz spike was an obvious indication of imbalance and/or misalignment and tear down the machine for balancing and realignment. After reassembly, the high-level 30-Hz peak may be reduced, only to return in a few weeks or months. Did the machine go out of balance again? What went wrong?

The novice failed to ask the previous necessary question: Is the level high because of an imbalance, or has a system natural frequency near 30 Hz been excited? The author knows of several consultants who make a good living by balancing the same machinery every month or two. However, balance is not the problem; there is a natural frequency in the system that coincides with 1 × rpm. As soon as some slight amount of erosion or chemical buildup occurs in the machine, the vibration level dramatically increases due to the instability of the resonance, and another balance job is performed. Instead the consultant should have found the cause of the resonance and recommended ways to shift the natural frequency away from 1 × running speed. A great deal more imbalance could then be tolerated before production would have to be halted for rebalancing.

Structural Response of a Simple Vibrator

A natural frequency is one at which the system exhibits very large magnitudes of vibration when excited by a very small force. Every structural system has infinitely many natural frequencies. When a system is wildly vibrating at its natural frequency, it is said to be in resonance.

In most cases, it is possible to consider a complex structure as the summation of all of its different resonances. One can therefore study the motion at any given natural frequency without considering the effect of other, widely spaced natural frequencies. Thus, a complex structure can be modeled as a set of simple spring-mass systems.

The curve of Figure 5.1 shows the response of a single-degree- of-freedom system. Such a system has the property that its motion can be described by only one parameter, such as the motion along the vertical axis on the spring-mass system of Figure 5.1A. This has been described as being the last vibratory system to be understood by man.

The curve has, for its vertical axis, the value of the resultant motion per unit of excitation force. The horizontal axis has the value of frequency divided by natural frequency. Thus, if one were interested in what happens to this system when excited by a force of 2 lb

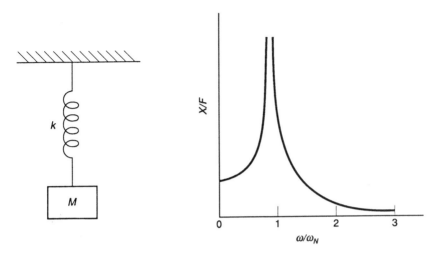

Figure 5.1. A single-degree-of-freedom system.

at a frequency of 3 × the natural frequency (note that the value of the natural frequency has not been specified), one would read the value of motion divided by force that corresponds to the horizontal value of 3 and double it (since the force of interest is 2 units in strength). One can also see from the phase plot the amount of lead or lag between the force and its response. It is obvious that if one has a forcing frequency near the natural frequency of the system, one can expect very large motions for a small excitation force.

The effect of the natural frequency must be taken into account when examining the vibration characteristics of a system. It could easily happen, for example, that a natural frequency of a pump–motor–piping–base system is equal to the blade frequency of the pump. If this occurs, the normally low-level blade forcing frequency could cause motion so large that couplings fail, bearings fail, or the structure actually begins to crack. It is therefore desirable that no significant forcing frequency existent in the pumping system be near the natural frequencies. It is sometimes necessary to change the mass or stiffness of the structure or the speed of the machine to eliminate a resonance problem.

In contrast, if one were designing a vibrating coal or rough-casting conveyor, the most economical frequency to operate at is a natural frequency, because smaller motors would be able to produce large motions. This is also true of cable-trench-making equipment which cuts through the ground at the blade/ground natural frequency.

The Importance of Structural Response: An Example

The nuclear regulatory agencies of various countries (including our own) insist that various pieces of critical machinery be capable of withstanding operation during an earthquake. One of the "standard" earthquakes is shown in Figure 5.2. The verification of this capability can be achieved either through a long series of analytical calculations (whose underlying assumptions can be thrown out at any time by the inspector assigned to review) or via an actual test on a shaker.

A given piece of well-designed machinery can usually pass the static g loading of the earthquake. The reason for failure is more likely due to the existence of a natural frequency in the region of frequencies where the earthquake has high energy. This has the effect of amplifying the earthquake at that frequency by a large factor. Thus, a machine without a natural frequency at, say, 1. 0 Hz will see a velocity excitation of 100 in/sec. A machine with a lightly damped natural frequency at 1.0 Hz might think that the same test is exciting it at 2,000 in/sec at 100 Hz. This machine will probably fail.

Effect of Damping on Structural Resonance

Figure 5.3A shows a representation of a single-degree-of-freedom system with damping. The damper is represented by a dash

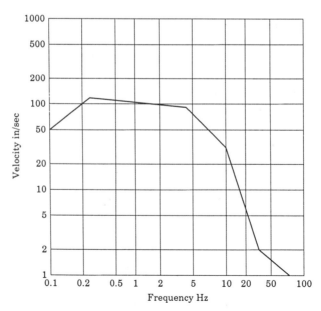

Figure 5.2. Earthquake Spectra.

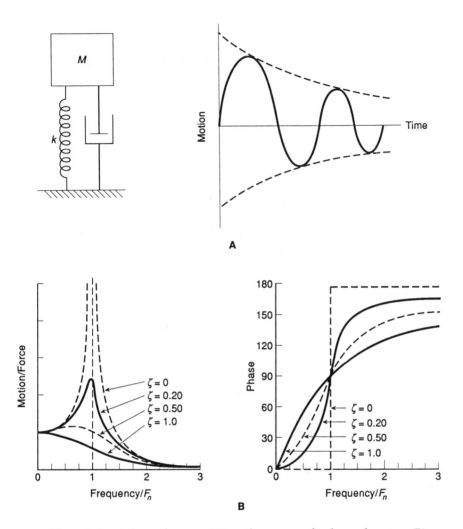

Figure 5.3. A damped system (A), and response of a damped system (B).

pot or, in automotive parlance, a shock absorber. The inclusion of damping in the model is an admission that, in the real world, energy is lost to friction during vibratory motion. This loss can be from the simple hydrodynamic friction of oil squirting through the small hole in the piston of a dashpot, from windage, or from other more complex frictional losses.

The amount of damping in a structure affects the resonance, such that, with no damping, one would expect to get infinite motion with very small forces. As the damping increases, the motion from a given force reduces. This is shown in Figure 5.3B, which is a plot of motion divided by force versus operating frequency divided by natural frequency for various levels of damping. The existence of damping also shifts the value of the natural frequency to a minor extent.

The Equations of Motion of a Single-Degree-of-Freedom System

The equations of motion of a single-degree-of-freedom system will be derived for the entertainment and edification of the math majors among us. Anyone else may skip this section.

The system of interest is shown in Figure 5.3A: a mass M, a spring of stiffness K, and a dashpot of damping coefficient B. The resultant force on the mass is

$$F = Ma = M \, dv/dt = M \, d^2x/dt^2$$

This force is balanced, as in Figure 5.4, by the spring force Kx and the damping force $Bv = B \, dx/dt$. Thus, a force balance yields

$$M \, d^2x/dt^2 + B \, dx/dt + Kx = 0$$

The classic solution to this differential equation is obtained by substituting $x = e^{st}$:

$$e^{st}(Ms^2 + Bs + k) = 0$$

This is obviously true only if the quantity in brackets equals zero. This yields a quadratic equation in s with two possible roots of s,

$$s_{1,2} = -B/2M +/- [(B/2M)^2 - K/M]^{1/2}$$

Figure 5.4. *A force balance.*

There are several things to note here:

- If there were no damping, one would plug in $B = 0$ above and find that the undamped natural frequency is

$$f = 1/2\pi \sqrt{K/M}$$

  where the 2π is included to convert from rad/sec to Hz.
- If $(B/2M)^2 = K/M$, the value of the square root part of the equation becomes zero and the resultant motion becomes an exponential decay with the shortest possible settling time. The value of B that causes this case is called *critical damping* and is denoted by B_c
- If $(B/2M)^2 >> K/M$, the system becomes *overdamped*. In this case, if the mass were excited by an impulse, it would ooze to a stop as if it were in molasses.
- If $(B/2M)^2 << K/M$, an *underdamped* system would exist. If the mass were excited with an impulse, it would oscillate sinusoidally for a rather long time.
- One can define a *damping ratio* as $\zeta = B/B_C$ such that $\zeta = 1$ for critical damping, $\zeta > 1$ for overdamping, and $\zeta < 1$ for underdamping. These are the values of damping shown in Figure 5.3B.

Notice that, in Figure 5.3B, if $\zeta = 0$ (undamped is a mathematical fiction), one gets infinite motion for zero input. This is an example of perpetual motion. If ζ is high, one could not even tell that the resonance exists except by observing a 90° phase change between the input and the output as the frequency approaches the natural frequency from either direction. Mechanical engineers have long used a term called the *log decrement* to describe the amount of damping in a system. It is given by the equation:

$$\delta = L_n (X_n/X_{n+1}).$$

Note that $\zeta = \delta/(4\pi^2 + \delta^2)^{1/2}$ (see Figure 5.5).

A term used by electronic engineers to describe filters is often used to describe the amount of damping in a system. The term is *amplification factor*, or Q. A high-Q system is one with little damping. A low-Q system is heavily damped. The relation between Q and ζ is $Q = 1/(2\zeta)$. The amplification factor is shown in Figure 5.6.

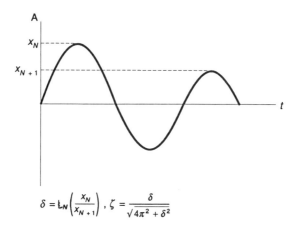

Figure 5.5. Log decrement.

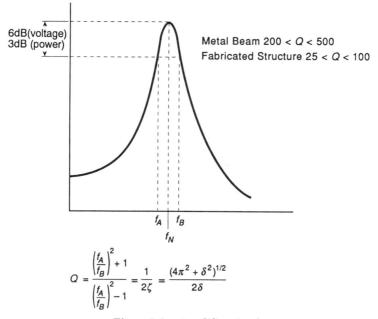

Figure 5.6. Amplification factor.

The amplification factor may be experimentally determined by measuring frequencies f_A and f_B, the half-power points of the response curve of the system. Note that the half-power points are 3dB, down from the peak response for power dB (10 log P/P_{ref}), or 6dB, down from (Voltage/Acceleration dB (20 log A/A_{ref}). The equation for Q is shown in Figure 5.6.

Estimating the Natural Frequencies of a System

Although calculating a system's natural frequencies is extremely tedious and subject to errors due to the accuracy of the mathematical model used, it often helps to decide how to shift an inconvenient resonance by thinking about the natural frequency of a single spring-mass system without damping. This was suggested to the author by the well-known A. H. Church of New York University in an undergraduate course on vibration. Over the years, it has probably proven to be the best suggestion for use during those times when the logical course of action is blurred by the urgency of the moment. The equation for this system, as shown above, is

$$f_N = 1/2\pi \sqrt{K/M}$$

where K is the spring stiffness in units such as pounds/inch and M is the mass of the system in units such as slugs.

Where it is necessary to raise a natural frequency to shift it away from a forcing frequency, either stiffen the supports or lighten the mass of the system. Softening the system or increasing the mass would lower the natural frequency. This runs counter to the intuition of many mechanics and maintenance managers.

Measuring the Natural Frequency

If the natural frequencies of a system are determined by measurement rather than calculation, one is freed from errors in the system's mathematical model (which may be almost impossible to derive). Sometimes, a natural frequency is assumed as a result of wacking the machine with a wooden 2" × 4" and measuring the resultant vibration with a single-channel spectrum analyzer. This causes two unanswerable questions:

1. If a peak in the acceleration response curve appears, is it there because a natural frequency was excited, or was this simply due to the spectral shape of the 2" × 4" blow?

2. If a peak fails to appear at a frequency where you expect a natural frequency to exist, is it missing because it doesn't exist or because it was not excited due to the frequency content of the 2" × 4" blow?

Although run up/coast down tests sometimes work on a slowly accelerating machine, there is no fully definitive way to measure a natural frequency using a single-channel spectrum analyzer. This is unfortunate, as a failure to decide whether a given excessively

108 *Chapter Five*

high amplitude vibration peak is due to the forcing mechanism or a resonance can result in many thousands of dollars wasted by fixing the wrong thing.

Two functions, transfer function and coherence, are required to properly do this job. These two-channel functions, as well as the third property of coherent output power, will be described below.

Cross-Channel Properties

Some people view a dual-channel spectrum analyzer as two single channel units in one box. Although there is some merit to this notion (i.e. monitoring costs can be cut almost in half because the technician can measure and record both the radial and axial directions of vibration at the same time), the power of dual-channel analysis will be completely lost in this kind of thinking.

There are three functions available only in multichannel analyzers that make dual-channel analysis extremely useful for certain classes of machinery-related problems. They are transfer function, coherence, and coherent output power. These functions will be explained in nonmathematical, practical terms in the following pages. Where mathematics may be useful to the mathematicians among us, it has been included in separate sections, which can be skipped by the rest of us.

Transfer Function — Mathematical Treatment

Although the transfer function is, in general terms, simply the complex ratio of sonic output divided by some input, most vibration specialists use the term to describe the complex ratio of a resultant motion divided by an exciting force. In the following mathematical description of a two-degree-of-freedom system (two masses), it will be seen that a transfer function defined as displacement divided by force turns out to be a useful quantity. If the reading of this section seems to progress too quickly for intellectual fulfillment, feel free to rederive the equation in terms of acceleration and force, or velocity and force. It works the same way. Feel free to skip to the next section if the math is of no interest.

Suppose we have the system shown in Figure 5.7, with 2 masses, 3 springs, and a force F acting on mass M_1. To develop the equations of motion, one need only do a force balance on each mass. Note from Chapter 3 that $x = X \sin \omega t$; $v = X\omega \cos \omega t$; and $a = -X\omega^2 \sin \omega t$.

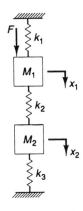

Figure 5.7. A 2-degree-of-freedom system.

Since the sum of the forces on $M_1 = 0$,

$$M_1 a_1 + K_1 x_1 + K_2(x_1 - x_2) = F$$

By the same token, the sum of the forces on $M_2 = 0$, such that

$$M_2 a_2 + k_3 x_3 - k_2(x_1 - x_2) = 0$$

Rewriting these equations after substituting the values for x and a and dividing by $\sin \omega t$ yields

$$-M_1 \omega^2 x_1 + k_1 x_1 + k_2 x_1 - k_2 x_2 = F$$

$$-M_2 \omega^2 x_2 + k_3 x_3 - k_2 x_1 + k_2 x_2 = 0$$

Therefore,

$$x_1 = \frac{F(k_2 + k_3 - M_2 \omega^2)}{(k_1 + k_2 - M_1 \omega^2)(k_2 + k_3 - M_2 \omega^2) - k_2^2}$$

and

$$x_2 = \frac{F k_2}{(k_1 + k_2 - M_1 \omega^2)(k_2 + k_3 - M_2 \omega^2) - k_2^2}$$

If we are interested in what the displacements x_1 and x_2 are for any given excitation force, we should divide the above equations by F. This is the displacement divided by force transfer function. If, at some time in the future, we had an interesting function for F, we could find the displacement of each mass by simply multiplying the two transfer-function equations by the function F.

$$TF_1 = \frac{x_1}{F} = \frac{k_2 + k_3 - M_2\omega^2}{(k_1 + k_2 - M_1\omega^2)(k_2 + k_3 - M_2\omega^2) - k_2^2}$$

$$TF_2 = \frac{x_2}{F} = \frac{k_2}{(k_1 + k_2 - M_1\omega^2)(k_2 + k_3 - M_2\omega^2) - k_2^2}$$

As a matter of further interest, the natural frequencies of this system can be easily calculated since natural frequency is, mathematically, the solution to the homogeneous (no forcing function, $F = 0$) differential equations of motion. Thus, if one rewrites the equations for x_1 or x_2 above with $F = 0$, it can be seen that the natural frequencies can be found by setting the denominator of either equation (they are the same) to zero and solving for ω (in radians/sec^2).

$$(k_1 + k_2 - M_1\omega^2)(k_2 + k_3 - M_2\omega^2) - k_2^2 = 0$$

$$\omega_n^2 = \frac{k_1 + k_2}{2M_1} + \frac{k_2 + k_3}{2M_2}$$

$$+/- \left[\left(\frac{k_1 + k_2}{2M_1} + \frac{k_1 + k_2}{2M_2} \right) - \frac{k_1 k_2 + k_2 k_3 + k_1 k_3}{M_1 M_2} \right]^{\frac{1}{2}}$$

If the above calculation looks painful, imagine what it would look like with the inclusion of damping and a few more degrees of freedom. Before the advent of computers, such problems took days to solve.

The more general case of a system with N masses is shown in Figure 5.8. One then must solve the equations using matrix methods or ascertain the natural frequencies by obtaining N transfer functions.

Dual-Channel Spectrum Analysis   111

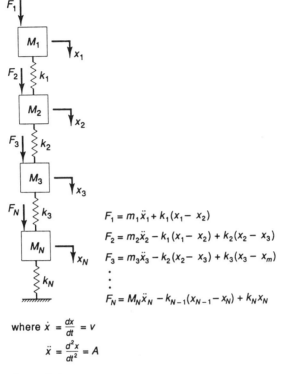

$$F_1 = m_1\ddot{x}_1 + k_1(x_1 - x_2)$$
$$F_2 = m_2\ddot{x}_2 - k_1(x_1 - x_2) + k_2(x_2 - x_3)$$
$$F_3 = m_3\ddot{x}_3 - k_2(x_2 - x_3) + k_3(x_3 - x_m)$$
$$\vdots$$
$$F_N = M_N\ddot{x}_N - k_{N-1}(x_{N-1} - x_N) + k_N x_N$$

where $\dot{x} = \dfrac{dx}{dt} = v$

$\ddot{x} = \dfrac{d^2x}{dt^2} = A$

In matrix form:

$$\begin{bmatrix} F_1 \\ F_2 \\ F_3 \\ \vdots \\ F_N \end{bmatrix} = \begin{bmatrix} m_1\omega^2 + k_1 & -k_2 & 0 & \cdots \\ -k_1 & m_1\omega^2 + k_1 + k_2 & -k_3 & \cdots \\ 0 & -k_2 & m_3\omega^2 + k_2 + k_3 & \\ \vdots & \vdots & \vdots & \\ 0 & 0 & 0 & \cdots \end{bmatrix} \begin{bmatrix} x_1 \\ x_2 \\ x_3 \\ \vdots \\ x_N \end{bmatrix}$$

$[F] = [A][X]$
The natural frequency is $[0] = [A]$. This happens when the value of ω is such that $|A| = 0$.

Figure 5.8. The general case of n degrees of freedom.

Transfer Function-Logical Treatment

The transfer function can be thought of as a system output (both magnitude and phase) divided by an input (magnitude and phase). The most common transfer function used in machinery analysis is motion (acceleration, velocity, or displacement) divided by excitation force or

112 Chapter Five

$$\text{Transfer function} = \frac{\text{acceleration} < \text{phase}}{\text{force} < \text{phase}}$$

One example of a simple transfer function is the natural frequency plot shown in Figure 5.3. Here, motion divided by force and phase difference are plotted versus frequency.

It turns out that, for linear systems, once the transfer function is obtained using any force (having an adequate frequency content), the motion can be predicted for any other force. That is,

New acceleration = new force × transfer function

Since it is possible to characterize a linear system by establishing the transfer function using any input force containing the frequencies of interest, a dual-channel-spectrum-analyzer test for transfer function can be carried out quite easily. In the simple impact test, a hammer with a force transducer is used as the excitation source (channel A) and an accelerometer is used as the output (channel B). If the point of impact and the accelerometer-mounting location are well chosen, the natural frequencies of the system can easily be determined by locating the peaks of the transfer-function magnitude plot or the point of 90° phase shift. Such a test (see Figure 5.9 for a schematic of the setup) would be conducted as follows:

1. Mount an accelerometer on the machine in question and connect it to channel B of a 2-channel analyzer.

2. Connect an instrumented force hammer to channel A.

3. Stop the machine. It is desirable that the only excitation to the machine come from the instrumented hammer blows.

4. Impact the machine with a few blows (it is surprising how a light blow with a small hammer can so easily excite a large machine). These blows will be needed to set up the analyzer for the test. This is often the most tedious part of the entire test.

5. Set the analyzer's trigger level (channel A). Set the input attenuation of channels A and B to avoid overload. Some analyzers will automatically change the weighting windows from Hanning to flat. If your analyzer doesn't, make the change at this time. Better analyzers will also ignore any data gathered during an overload condition.

Some analyzers allow for pre-screening of each impact event before inclusion into the averaged result.

6. Choose a time window (1/bandwidth) which shows the proper "ring down" of the time domain output of the system (see Figure 5.10).

7. Average several blows. View the time domain signals of both channels during this test to assure yourself that the blows being imparted are good clean blows. Some analyzers allow the use of special weighting windows to help assure this.

8. View transfer-function magnitude and phase. The natural frequencies of the system are those points where the magnitude versus frequency plot peaks and the phase versus frequency plot shows a phase shift of approximately 90° for the force gauge/accelerometer instrumentation described here. See Figure 5.11 where natural frequencies of 21.7 Hz and 387.5 Hz have been found.

9. View the coherence display (described below). A coherence value near 1.0 indicates a good test. Where the value of coherence is low, something invalidated the test, and the transfer function should be ignored at these frequencies. See the following section on coherence for a discussion on some of the possible errors in testing that could cause invalid results. Figure 5.12 shows the coherence and coherent output power versus frequency plots for the test described.

10. The values of the damping ratio at the natural frequencies just determined can be calculated by

$$Q = \frac{(Fa/Fb)^2 + 1}{(Fa/Fb)^2 - 1} = \frac{1}{2\zeta}$$

where Fa and Fb are found as in Figure 5.6.

Note that, by using transfer function, both the system's natural frequencies and the damping factors can accurately be measured. The coherence at these frequencies indicates the level of confidence with which conclusions can be drawn.

114 *Chapter Five*

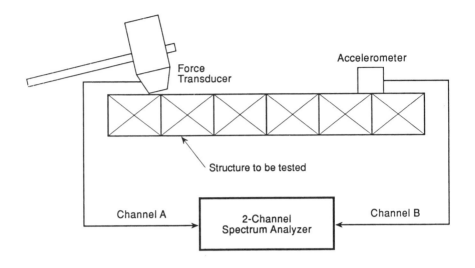

Figure 5.9. Schematic of a simple 2-channel impact test.

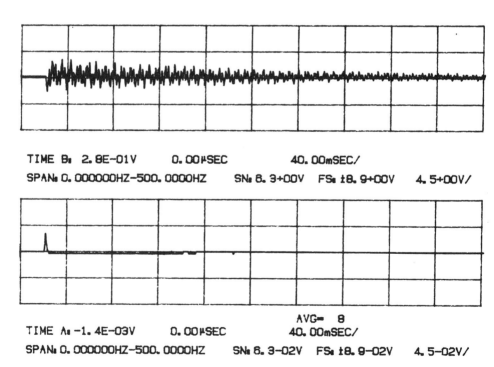

Figure 5.10. The time window of the analyzer should be chosen for a proper ring down of the motion (above). The lower time history shows the force.

Dual-Channel Spectrum Analysis 115

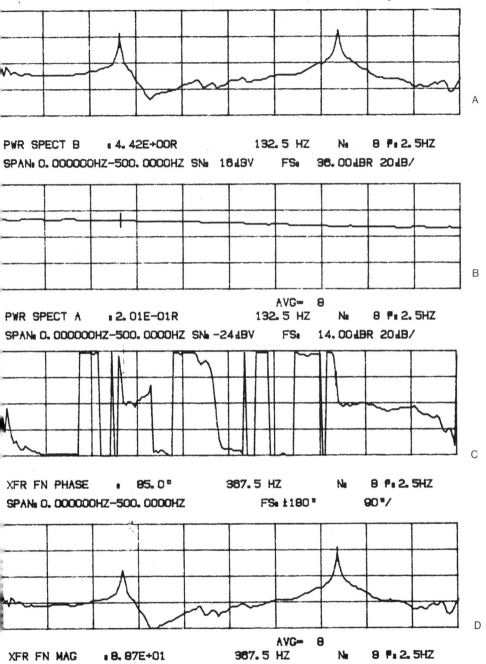

Figure 5.11. A. The power spectrum of the motion. B. The power spectrum of the force. C. The transfer function phase shifts (ϕ hammer − ϕ accel.). D. The transfer function magnitude versus frequency.

116 Chapter Five

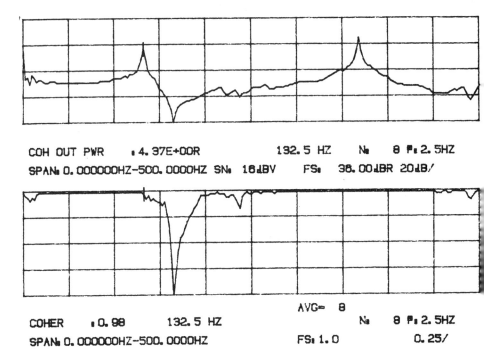

Figure 5.12. *The coherence for the data of Figure 5.11 shows a problem in the results in the region about 160 Hz. Elsewhere, the test was good.*

Other Kinds of Inputs for Transfer-Function Testing

It is possible to test for natural frequencies in several ways other then striking the structure with an instrumented hammer. One can excite the structure with an oscillatory motion using an electromagnetic or pneumatic shaker. This requires the ability to vary the frequency of the oscillatory motion of the shaker over the frequency range of interest, and usually requires a rather expensive shaker and control system. Some multichannel spectrum analyzers are capable of supplying several different kinds of excitation to the control system of the shaker. A few of these are described below:

White Noise This input, which can be approximated by an impact hammer blow, contains equal energy at all frequencies. The advantage is that all frequencies can be excited at once. The disadvantage is that, since the energy is distributed over all frequencies, there is not much energy at any one frequency and it may be impossible to adequately excite a particular damped resonance due to the inability to impart enough energy at all frequencies. It is conceivable, in fact, to wreck the system under test because the energy necessary to excite one frequency can be enough to overload the system at some other frequency.

Band-Limited White Noise Unlike white noise, this kind of excitation has equal energy only at all those frequencies contained in the analysis range of the spectrum analyzer.

Pink Noise Pink noise was a popular form of excitation when constant-percent tunable filters were used. Since a low-frequency constant-percent filter is narrower then a high-frequency filter, it required less energy in its filter band to read a particular level. Pink noise takes account of this by supplying lower amplitude excitation at high frequencies than at low frequencies. On a narrow band spectrum analyzer, pink noise looks like a downward ramp.

Sine Sweep A sine wave, or several adjacent frequency sine waves, are swept through the frequency range of interest, eventually exciting all of the frequencies in the analysis range. This method has the advantage that all of the input energy resides at one particular frequency at any given instant and this level can be adjusted to optimize the dynamic range of the test. The disadvantage is the time it takes to sweep through a frequency range. Because of the swept character of this kind of test, tunable filter devices are somewhat better suited to this test than FFT spectrum analyzers.

Chirp There are various kinds of chirp signals that can be input to the shaker, and are comprised of short bursts of sine sweeps. These chirps allow for a more rapid test than sine sweeps as a series of averages can be taken with a spectrum analyzer over its analysis range rather than requiring the sine sweep to dwell at each frequency.

Keep in mind that any of the above excitations of a shaker is distorted by the transfer function of the shaker itself, as well as the way in which the system under test and the force transducer are affixed to the shaker. For this reason, shaker testing is an art in itself, and coherence should be used to validate the testing.

Coherence Explained

Coherence is, basically, a cause-and-effect parameter. It can only be implemented in a multichannel spectrum analyzer because it is necessary to simultaneously gather both the cause signal and the

118 *Chapter Five*

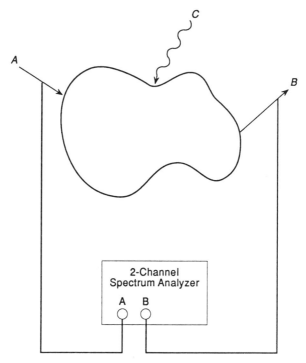

Figure 5.13.   Test of a linear system.

effect signal. Thus, if the coherence between two channels has a value of 1.0, all of the signal of one channel was caused by the signal of the other channel in a linear fashion.

For example, for the test of a typical linear system, as shown in Figure 5.13,

- Coherence = 1.0 indicates all of the output to the effect channel (B) was due to the input at the cause channel (A).

- Coherence = 0.0 indicates none of the effect at B was due to the cause at A.

- Coherence = 0.5 indicates that 50% of the output at B was caused by the input at A. If input C was the only remaining input to the system, the coherence between C and B would, in this case, also be 0.5.

Some Uses of Coherence

There are several classic uses for the coherence function in machinery analysis. Three of these will be discussed:

Transfer-Function Check After obtaining a transfer function between two points on a machine (as described in the section on transfer function), the correctness of the measured results must be determined. Several sources of error exist:

- Nonlinearity of the system under test
- Relatively high noise input levels that effect the output signal of the system but are not part of the measured input signal (see the section on inverse transfer function in Chapter 8)
- Other system inputs that were not part of the measured input signal but did effect the output signal
- For impact tests, the failure to wait until the results of the first impact settles out before applying a successive impact (in this case, the residual of the first impact acts as an incoherent input during the second blow)
- Insufficient averaging
- System time delays that are large compared to the time window (1/bandwidth) chosen for the test
- Attenuation settings that do not take advantage of the full dynamic range of the analyzer, causing the internal noise of the analyzer to interfere with one or both of the measured signals at frequencies where low energy exists

It is extremely risky to draw conclusions from a transfer-function test without checking the coherence display for an indication of one of the above errors. A low value of coherence at a frequency of interest should prompt a reevaluation of the testing procedure. Many spectrum analyzers have the ability to blank out those segments of a transfer-function display that have a coherence below a certain preset level.

The plot of Figure 5.14 shows both the transfer-function magnitude of a simple system and the associated coherence. Note that the coherence is low above and below the system resonance (where the system noise was large compared to the measured signals).

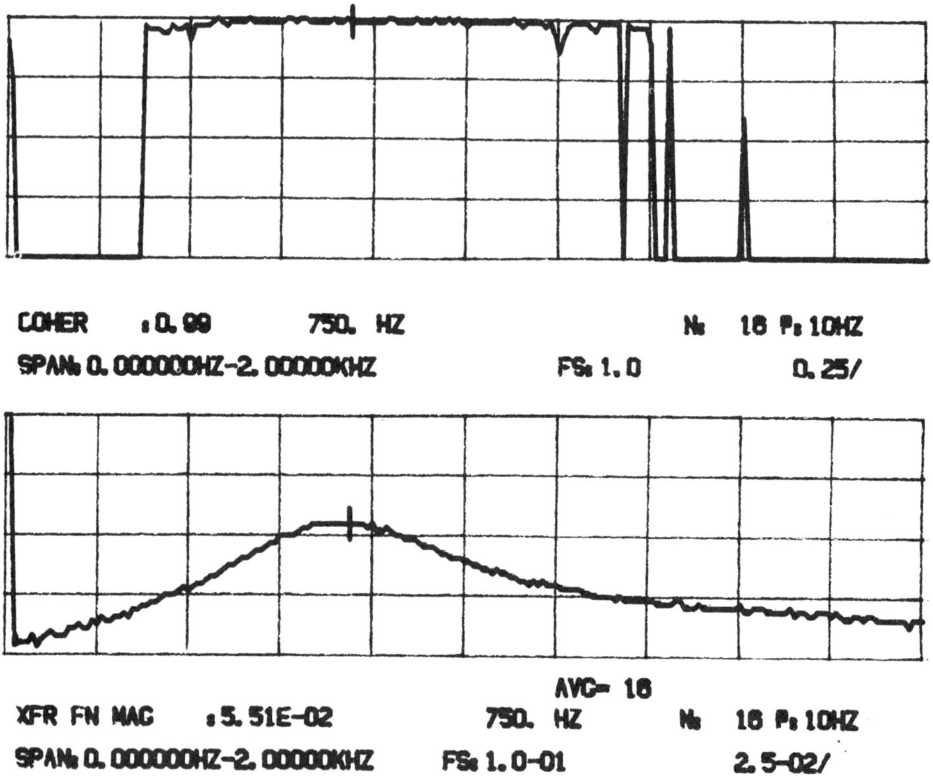

120   Chapter Five

```
COHER    : 0.99      750. HZ                    N: 16  P: 10HZ
SPAN: 0.000000HZ-2.00000KHZ        FS: 1.0              0.25/
```

```
                                     AVG= 16
XFR FN MAG    : 5.51E-02    750. HZ              N: 16  P: 10HZ
SPAN: 0.000000HZ-2.00000KHZ    FS: 1.0-01            2.5-02/
```

Figure 5.14. Plot of coherence and the magnitude of a transfer function. Note that the coherence drops off above and below the natural frequency because the output of the system is masked by internal analyzer noise due to improper attenuator settings.

Accelerometer—Magnet Check Often, the question of the frequency response of an accelerometer-magnet combination puts certain conclusions about the higher frequency vibration of a machine in question. For example, one might ask if a 4-kHz peak in the spectra of a gearbox is the correct amplitude of a mesh frequency or the result of a resonating magnetic mount. A simple way to answer this question is to mount a second accelerometer set next to the first and check their coherence on a 2-channel analyzer. If the coherence at 4 kHz is high, accelerometers are sensing the same vibration and, by implication, the magnet is properly holding its accelerometer to the surface. If the coherence is poor, a mounting block will have to be glued to the surface of the gearbox for accurate readings.

Source of Vibration Check The source question has two common origins. In one case, it might be desired to determine which of several machines are causing excessive noise in a work area. The second case is that in which it must be determined if a given machine is vibrating excessively due to its own forcing mechanisms and resonances or whether it is being shaken by an adjacent machine (neither of which can be turned off to run the obvious single-channel background tests). Since both of these problems require the same technique, only the former will be discussed in some detail:

Suppose three similar machines, machine 1, machine 2, and machine 3 are operating near a work area in which the sound level at the ear of the listener is 95 dBA. The coherence function can be used to determine which machine (or machines) is causing the problem.

1. Connect a microphone to channel B of the spectrum analyzer and place it in the work area
2. Connect an accelerometer to channel A of the analyzer and mount it on machine 1
3. Measure the coherence (machine 1 only causes a high noise level at those frequencies where the coherence is high)
4. Repeat the above measurement with the accelerometer on machine 2 and again on machine 3
5. Compare the three coherence plots; the machine with the highest coherence values over the widest frequency range is the largest contributor of sound energy at the microphone

Note that, in the above tests, it was not necessary to stop any of the operating machinery. A direct reading of the sound power at the microphone due to any given machine (as if the other machines were not working) can be had by looking at the coherent-output-power function described below.

Coherent Output Power

A function available on most 2-channel spectrum analyzers is coherent output power, which is closely related to coherence. It is the power spectrum at channel B of the analyzer multiplied, at each frequency, by the coherence between channels A and B.

Thus, in the above example, the coherent output power measured while the accelerometer was mounted on machine 1 corresponds to the sound level one would measure at the microphone if only machine 1 was operating. This result was obtained even though all of the machinery was running during the test.

Note that this is somewhat different than trying to determine the noise due to machine 1 by synchronous time-averaging. In the latter case, only the noise synchronous to the shaft of machine 1 will be measured. The noise from all non-synchronous noise sources of machine 1 will not appear. The analyst will likely blame these noises on the other two machines or on background noises rather than asynchonous sources in the guilty machine.

Modal Analysis

To determine the vibration characteristics of a complicated device, such as an airplane or a truck, it is possible to do a modal analysis to obtain the natural frequencies, damping coefficients, and mode shapes of the structure. Mode shapes are the deflection shapes taken on by a system when excited at one of its natural frequencies. As a simple example of what mode shapes look like, observe some of the mode shapes of various beams excited at some of its natural frequencies, as shown in Figure 5.15.

To do a modal analysis, follow these steps:

1. Model the system by breaking it up into a set of lumped masses connected by a set of spring/damper elements. If a finite element analysis were being performed, it would be necessary to know the values of these masses, springs, and dampers. In a modal analysis, the masses are used as measurement points and the properties of the structure are determined experimentally.

2. Measure the transfer function between each of the mass points. One way to do this, for instance, would be to always impact point 1 and measure acceleration at all of the points in the structure. This would yield $TF_{1,1}$, $TF_{1,2}$, $TP_{1,3}$, etc.

3. Given the transfer functions, it is obviously possible to determine the system's natural frequencies and effective damping. To determine the mode shape at any given frequency, one need only assume a deflection of one unit at point 1 and use displacement transfer functions to determine how all of the other points will move relative to point 1.

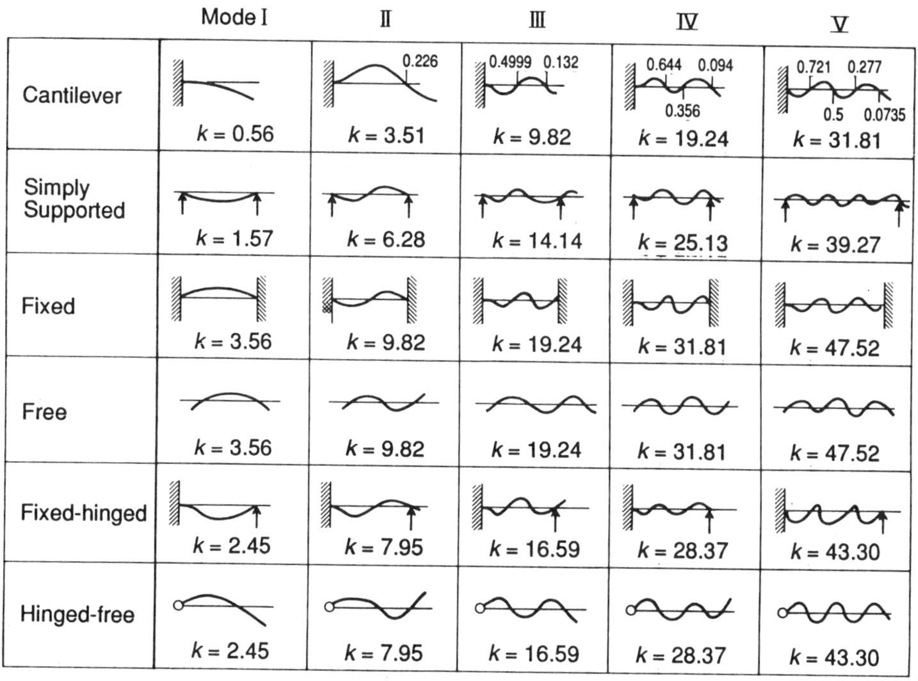

	Mode I	II	III	IV	V
Cantilever	$k = 0.56$	0.226 $k = 3.51$	0.4999 0.132 $k = 9.82$	0.644 0.094 0.356 $k = 19.24$	0.721 0.277 0.5 0.0735 $k = 31.81$
Simply Supported	$k = 1.57$	$k = 6.28$	$k = 14.14$	$k = 25.13$	$k = 39.27$
Fixed	$k = 3.56$	$k = 9.82$	$k = 19.24$	$k = 31.81$	$k = 47.52$
Free	$k = 3.56$	$k = 9.82$	$k = 19.24$	$k = 31.81$	$k = 47.52$
Fixed-hinged	$k = 2.45$	$k = 7.95$	$k = 16.59$	$k = 28.37$	$k = 43.30$
Hinged-free	$k = 2.45$	$k = 7.95$	$k = 16.59$	$k = 28.37$	$k = 43.30$

$$F = k \sqrt{\frac{gEI}{wL^4}}$$

where F = natural frequency, Hz
g = 386 in/sec^2
k = constant shown in figure

E = modals of elasticity, lb/in^2
I = moment of inertia, in^4
w = lb/in
L = beam length, in

Figure 5.15. Beam natural frequencies (adapted from Church: Mechanical Vibrations, 2nd Edition).

Although it is possible to do a modal analysis using a 2-channel analyzer, a tremendous amount of point picking from the transfer function plots and a large number of hand calculations are required. Most people simply buy a modal analyzer. This is nothing more that a multichannel spectrum analyzer tied to a computer for the number crunching.

A good modal analyzer should have the following capabilities:

1. An easy way to enter the geometry of the system—such simplifications as symmetry and repetitive subsets of points should be available to reduce the number of points that must be manually entered into the computer; the program should be able to handle a large number of points

2. A quality multichannel analyzer capable of zoom for closely spaced natural frequencies—some modal systems use a dual channel analyzer, more expensive systems use a 4-channel analyzer, one channel for the input and three channels for the three mutually perpendicular axes of the response at any point

3. The ability to handle complex roots of the equations—a few modal systems have been built that simply ignore nonreal eigenvectors, which can result in severe errors

4. Several different curve fitting algorithms—remember that, although humans can visually look at the peaks of a transfer-function curve to find a natural frequency, a computer must do curve fitting to learn what the transfer function looks like in the digital world; this is particularly important for closely spaced roots

5. For added convenience, the ability to show a good animated display of the way in which the system (or parts of the system) moves at any given frequency and the ability to rotate the display of the system on the screen to better facilitate viewing of some particular part of the modeled system

6. More advanced modal systems allow the user to take the model that was tested and add different components, such as new masses, more bracing, and so on, mathematically rather than physically, which allows what-if analysis on a system with which the user is not completely pleased without welding new components to the system and retesting

Case Histories

A Power Plant Problem

A new power plant had two large pump packages mounted on a mezzanine, as shown in Figure 5.16. Unit #2 ran well, but unit #1 shook the entire mezzanine. The pump manufacturer was being blamed. The threat of multiple lawsuits filled the air. The power company was refusing to accept delivery of the plant until the problem was resolved. The engineering company and construction company were ready to sue each other as well as the pump manufacturer.

Dual-Channel Spectrum Analysis 125

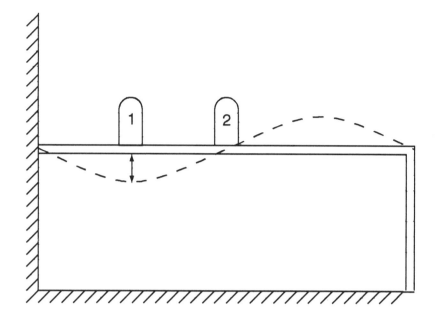

Figure 5.16. Two pumps on a mezzanine.

A set of tests were run by the author using a simple tunable filter vibration meter. (Some of us are old enough to pre-date FFT equipment.) Both pumps were shut down. A pneumatic impact hammer was fastened to the mezzanine near pump #2. The speed of the hammer was slowly adjusted until everyone present agreed that their feet tickled; most tunable filter analyzers are too slow to run "real time" data as can the modern FFT data gathering box. Vibration data taken on the pneumatic impact hammer showed that the predominant frequency excited by the hammer was equal to the blade frequency of the pump.

Solution: The conclusion was simple. The design of the mezzanine was such that it had a natural frequency equal to the blade frequency of the pumps. Pump #2 was located near a node of the floor, making it almost impossible to excite the structure at that frequency from that location. Pump #1 was at an anti-node, making excitation at that location very simple. The structural design engineer was at fault. The pump manufacturer, through the use of simple filtered vibration equipment, was found to be innocent.

Petro-Chemical Plant Problem

Construction of a billion-and-a-half-dollar plant in Europe was being held up by the inability of a pump supplier to build a pair of compressor packages which would meet the vibration criteria of this Fortune 500 oil company. The equipment manufacturer contended that he had done all he could to reduce the vibration and that the specs were unrealistic. GMDC was called in by the oil company to evaluate the problem.

Solution: After several meetings with the client and the manufacturer, Transfer Functions determined that the bearing arrangement of the equipment had a natural frequency which corresponded to one of the forcing mechanisms at 50 Hz operation. Because the manufacturer refused to redesign the compressor, a frequency converter had to be employed to allow the compressor to operate at a 52 Hz speed. Not coincidentally, a similar order from the client for the same equipment at a Texas site was changed to specify a gear ratio change corresponding to a 52 Hz operating speed.

Paper Mill Speed-Up Problem

The newest paper machine in one of the largest paper-making facilities in the world was scheduled to be sped up by about 10%. Upon approaching the new operating speed, one of the major drying rolls exhibited excessive vibration, forcing a slow-down. The potential loss of product was projected in the millions of dollars.

Solution: The new operating speed caused one of the forcing frequencies of the troublesome roll to excite a pre-existing structural resonance. A redesign of the roll assembly to reduce the amplitude of the excitation, plus a change in nip pressure to slightly shift the natural frequency of the system, has allowed operation at the new speed until more extensive changes could be made.

Solid State Electronics Problem

A leading manufacturer of solid state electronics devices moved into a new plant in south Florida. After several months, they were still unable to grow crystals at a commercial rate. Further, the inspectors, whose job it was to inspect chips under microscopes, were spending vast amounts of time nauseous in the restroom. It was suspected that some form of vibratory excitation was disturbing the crystal growth.

Solution: Through the use of the dual channel spectrum analysis functions known as "coherence" and "coherence output power," it was determined that a relatively small reciprocating compressor in the utility room was shaking the entire foundation of the plant at 10 Hz. Not only was this vibration disturbing the crystal growing tanks, but it was also causing small vibrations at the microscope tables, which was amplified when looking through the microscopes, causing the nausea of the inspectors. The compressor was isolated. Chip growth and restroom usage returned to normal.

Summary

The dual-channel spectrum analyzer can now be used in actual plant environments. This means that transfer function and coherence can easily be used to avoid the guess work and resultant errors in problem diagnosis that cost industry untold amounts of money in down time, machine failure, and lost production.

With a better understanding of 2-channel analysis capability, plant machine-reliability personnel should be able to do a better job of problem diagnosis than has heretofore been possible. This improved ability can only serve to save money which would otherwise be wasted fixing problems that don't exist and ignoring problems that are likely to be recurrent.

Finally, it was shown that a modal analysis system was little more than a multichannel spectrum analyzer and a computer. Its advantage is in the improved ability of the user to visualize the motions of complex structures and mathematically modify them for performing hypothetical what-if studies.

CHAPTER SIX

Periodic-Condition Monitoring

Introduction

This chapter deals with the art of periodically monitoring the condition of various important pieces of rotating machinery in a plant to determine the present health of the equipment. Another common term for this endeavor is *predictive maintenance*. The reasons for carrying out such a program are given below:

Eliminate Unnecessary Disassembly Any time a machine is tampered with, there is a possibility that it will be damaged. Under the old philosophy of preventative maintenance, for instance, the bearings might be replaced every 2 years whether they needed to be or not. This led to a real possibility that

- The new bearing would fail in a short time due to manufacturing errors in the bearing itself (infant mortality) or due to improper installation
- A lock washer, lock nut, or lock wire might be forgotten, causing the machine to fall apart after just a few hours of operation
- Poor assembly clearances could cause rubs, premature wear, or operating inefficiency
- Poor alignment of the rebuilt machine or improper lubrication could cause bearing or coupling failure in a short time

- A forgotten part of the machine or a tool could be sucked into the inlet of either the rebuilt machine or another machine in the system, wrecking it.

These kinds of experiences led to the sage advise, "If it ain't broken, don't fix it." A machine that is exhibiting no sign of wear or faulty operation should be left alone to do its job for as long as possible.

Reduce Unscheduled Downtime If a particular machine is beginning to exhibit signs of incipient failure, it is possible to schedule its repair during a convenient time (i.e., during a planned shutdown). Provision can be made to ensure that the proper replacement parts, tools, equipment, and manpower will be available at the appointed time. This kind of planning will obviously reduce the annual maintenance budget and help ensure that the repair will be completed on time. In this way, the minimum amount of lost production time will be experienced.

Avoid Wrecks By following the condition of each machine as it ages, one can predict the onset of many destructive modes of failure before they occur. By not running a machine to total destruction, the chances of making a rapid repair are greatly enhanced, and the probability of injury or death to operating personnel, who might otherwise happen to be near the machine during failure, is removed.

Reduce Insurance Costs Two types of insurance can possibly be reduced by demonstrating that a plant's monitoring system can successfully reduce machine failure:

- Lost production insurance, which reimburses a company for its inability to produce its product due to equipment failure
- Liability insurance, which protects the company from lawsuits resulting from death or injury

The Goals of Predictive Maintenance

It is almost impossible for a machinery monitoring program to be cost ineffective. A very simple cost-effective program could be devised as follows:

Drill a hole in the night watchman's shoe. Instruct him to walk past each major machine in the plant each night. Have him write down the machine number of each machine near which his foot gets wet.

Such a simple-minded monitoring program will be cost effective because, aside from costing almost nothing (the watchman is being paid to walk around anyway), maintenance personnel will get a daily list of all the failed mechanical seals, all the loose packing, piping leaks, and an indication of which bearings may have had their grease washed out by water leakage.

Thus, by simply assigning a person to stand next to a piece of machinery on a regular basis, large sums of money may be saved by locating certain potentially expensive problems. The simple goal of cost effectiveness is not enough of a reason for following a preventive maintenance program. The goal should be fewer unexpected failures in the plant, maximum running time between shut downs, and the elimination of recurrent failures.

Furthermore, given the fact that any monitoring program will show a cost savings by pointing out some potential machine failures, the challenge is to design a program that will improve the success rate of the program with minimum capital investment. Simply put, the success rate can be thought of as the probability of finding a problem that exists and of not finding a problem that doesn't exist (as this kind of error can be costly, too).

Relevant Terminology

Many people interested in the advantages of predictive monitoring discussed above fail to properly delineate the different aspects of monitoring and begin setting up a program using the wrong type of equipment, improper instrumentation, poor data, and improper job descriptions for the personnel assigned to do the actual work. In recent years, sales hype by the manufacturers of "monitoring boxes" has further muddled the picture. Therefore, several terms should be discussed at this point in order to avoid future confusion.

Continuous Monitoring

One continuously monitors a critical piece of machinery by permanently mounting a number of transducers (usually temperature probes, proximeters, accelerometers, etc.) on the machine at the proper places. Each pickup is permanently wired to a panel meter in the control room. The panel meters have a set point with which to use the incoming signal to trigger a warning alarm. A higher set point at which either a higher level alarm goes off or the machine shuts down is also provided.

Very often there will be some sort of voting logic in the decision to shut down. An example of this is in the case of monitoring the thrust bearing on a steam turbine: A proximeter is mounted such that it "looks" across the oil film between the thrust-carrying surfaces of the bearing. In addition, one or more temperature sensors are mounted on the thrust pads. If the gap seen by the proximeter approaches zero, it could mean that the sensor is broken. If the thrust pad temperature increases, it could mean a problem in the thermal couple system. If, on the other hand, the oil film gap goes to zero as the temperature rises, both pickups are probably working and the turbine is about to crash. The voting logic of the monitoring system would shut the machine down.

Although continuous systems are designed to look at every measured parameter during every second of machine operation, the data gathered is usually in an unrefined, unfiltered state. Thus, such a system will not tend to see incipient failures of low-energy mechanisms, such as bearings or couplings, until the failure is imminent. A periodic monitoring system using filtered vibration is required for advance notice of such problems.

Because of the cost of mounting a great many sensors and wiring them into a control room, continuous monitoring tends to be quite expensive. The technique is therefore used sparingly.

Periodic Monitoring

Originally, this technique was used only in place of continuous monitoring for those machines where the cost of a continuous monitoring system was not justified. The method was only one step better than the previously described watchman-with-a-hole-in-his-shoe method. In this case, the watchman carried an overall velocity-reading meter from machine to machine on a weekly or monthly basis. The levels at each bearing of each machine were written down in the hope of finding some incipient failure.

One of the problems with this simple method of operation is that, while it is cost effective, some types of failure are unlikely to be predicted. Certain forcing mechanisms have high amplitudes of vibration at their forcing frequencies when they are operating normally, whereas other mechanisms do not exhibit significant levels

132   Chapter Six

until the instant of failure. A typical example of a machine having both high- and low-forcing levels is a simple gearbox.

For example, take a gearbox having, among other things, a mesh frequency of 1 kHz and an outer-race-ball pass frequency (rolling element bearing) of 236 Hz. If the vibration was monitored with an overall meter, it would be impossible to separate the two signals (their energies would add). Since the gear mesh level is high compared to the bearing level, any changes in the vibration level of the bearing due to incipient bearing failure will be masked by the normal energy output of the gear mesh.

If a measuring device using several filters was used to monitor the gearbox, the bearings and gears could be monitored separately. Table 6.1 shows possible level changes in the gearbox vibration due to a failing bearing. Note that while the bearing problem is evident in the filtered data, it never shows up in the overall data because it is being hidden by the high-energy content at the mesh frequency.

The bearing failure and consequent destruction of the gearbox would have come as a complete surprise to a technician monitoring the gearbox with an overall meter. A technician with an FFT-based monitoring device, on the other hand, probably would have saved the plant enough money by avoiding this one failure to pay for several monitoring boxes. Figure 6.1 shows a failing bearing in a paper machine dryer in the presence of the gear mesh. An overall meter would not have found this problem.

Table 6.1.

Time	Overall Levels, g's	Gear Mesh, g's	Bearing Outer Race Fault Frequency	Comments
Month 1	6 g's rms	6 g's rms	.002 g's rms	Everything OK
Month 2	6 g's rms	6 g's rms	.004 g's rms	Incipient Bearing Failure
Month 3	6 g's rms	6 g's rms	.006 g's rms	Imminent Bearing Failure

The most commonly masked forcing frequencies are from bearings and worn couplings. The signals likely to mask them are unbalance, blade frequencies, and mesh frequencies. For this reason, many modern periodic monitoring endeavors use filtered vibration readings to enhance the ability to predict incipient failure—hence the term predictive maintenance.

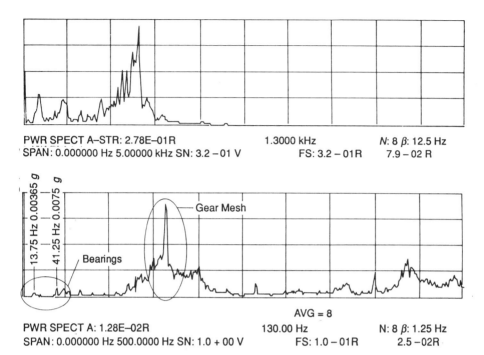

Figure 6.1. Bearings on a paper machine dryer in the presence of a gear mesh.

Trending

Although the simple comparison of currently gathered filtered vibration data to a predetermined baseline will generally indicate a problem in a machine, the ability to estimate how much longer the machine may be allowed to run before repair requires a long-term trend of the machine's condition with time. For example, a comparison of the vibration levels of two different bearings, with about the same baseline level and current level, will show that bearing A is slowly deteriorating while bearing B is about to self-destruct (see Figure 6.2).

Methods for archiving data for trending, suggested methods of displaying trended data, and methods of reducing the amount of data to be stored will be discussed in later sections of this chapter.

Untrendable Failures

An untrendable failure progresses from its initiation as a barely detectable fault to a complete failure in a time period that is shorter than the periodic monitoring time interval assigned to the machine. Three examples of untrendable failure are as follows:

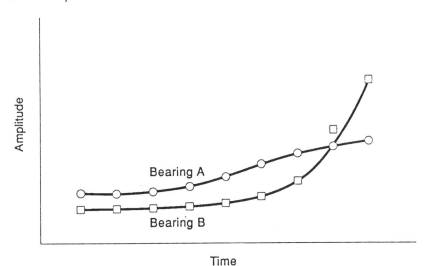

Figure 6.2. Trending the deterioration of two bearings.

1. A rolling element bearing that fails from fatigue in 3 months when the periodic monitoring interval for the particular machine is once every 6 months

2. An oil film bearing that has lost its lubricant when a continuous monitoring program is not being used

3. A wrench sucked into the inlet of a pump. Said wrench having been left in the line by a careless mechanic or a saboteur

Diagnostics

One diagnoses a problem to get a clear understanding of its nature. In this way, optimal decisions as to the cure of the probable problem can be found. The art of properly diagnosing a problem usually requires an expert with a great deal of experience and the use of very sophisticated equipment. Some of the techniques of diagnosis were discussed in Chapters 4 and 5.

Because of the cost of the equipment and the amount of experience required to diagnose difficult problems correctly, the diagnostician tends to be a different person then the monitoring technician. Smaller companies tend to use in-house personnel for monitoring and trending but call in an outside consultant for critical diagnosis of important problems. Large corporations often have a monitoring group in each plant, which is supported by a group of experts located at corporate headquarters.

Decisions to be Made When Getting Started

The ability to find as many impending machinery failures as possible (without deactivating or opening the machine unnecessarily) requires a system of gathering data for comparison to set the limits, trending the data over long periods of time, and the careful selection and location of transducers. If a predictive periodic monitoring program is to meet the goals mentioned above, considerable thought must go into the initial design of the program. If the instrumentation and implementation of the program is left to chance, the goals will not be met.

Considerable thought must be given to the philosophy of a machinery health monitoring program before instituting one. Which machines should be monitored and how often? How is the training of personnel to be carried out? The answers to these questions are very much a function of the individual plant and the complexity of the machinery involved.

A program so complex in design that it will take years to put into operation is of little present value because the monitoring function will not be set in motion for too long a time period, permitting a large number of interim machine failures. Such grandiose programs tend to lose the support of management and the interest of the staff long before they ever get off the ground.

Although it is beneficial to gather data on a piece of machinery and know that, for example, a particular bearing is failing for the third time in a year, it is certainly more useful to know why the bearing regularly fails. To accomplish this goal, the monitoring program must have at least enough sophistication to provide a diagnostician with enough historical data to correlate the causes and effects of failure for any particular machine.

The issues that must be decided before the first piece of monitoring equipment is selected or the first technician is assigned to the monitoring function will be now discussed. If the reader's plant already has a monitoring program, these issues can be used to evaluate the program and to decide how to improve its shortcomings.

Which Machines Should be Monitored?

This question must be considered at the outset. Continuous monitoring programs generally involve machines that meet one of the following three criteria:

1. The machine is critical to production
2. The machine is extremely costly to replace
3. Failure of the machine could easily result in injury or loss of life

The criteria for continuously monitoring a machine are stringent because the cost of using dedicated transducers hard wired to a control room is quite high. Although a good accelerometer can be bought for $300–400 and panel meters are also generally less then $1000/channel, the cost of running cable and maintaining the system is often several hundred thousand dollars for enough channels of data to adequately monitor the machinery of interest.

Furthermore, note that the system usually installed does not use filtered vibration. Thus, the gradual failure of a low-energy component, such as a bearing or a coupling, is not likely to be noticed until the situation is critical and considerable damage is about to occur. The chances of an orderly, planned shutdown are not particularly likely in this instance.

Some rather exotic computer-based programs have been added to extremely crucial machine trains. These systems are based on the usual continuous monitoring scheme with one important difference:

The vibration signal, which is normally connected to a panel meter, is instead sent to a multiplexer. The multiplexer scans each channel of data every 15 min–1 hr. This data is fed to an FFT spectrum analyzer and then to a computer. The computer compares the spectral data to a preestablished baseline for that particular channel and decides whether the data has exceeded the baseline at any frequency. If the baseline is exceeded, an exception report is issued. If not, the data is disregarded (to avoid overloading the data retention capabilities of the computer), and the multiplexer switches to the next channel of data.

The advantage of this obviously expensive system is that one has the usual advantages of continuous monitoring (the machine is monitored at every second of the day and untrendable failures cannot "sneak past" the monitoring system), as well as the periodic advantage of gaining knowledge of incipient failures of low-energy devices in time to take preventive action.

Periodic monitoring programs are sometimes carried out on machines meeting the above criteria, but more often include the following, less critical criteria:

1. The machine is fairly costly to repair or replace
2. Failure of the machine may cause increased production costs or reduced plant output, but not necessarily a complete plant shut down
3. Machines with a poor operating history

The machines meeting these criteria seldom justify spending hundreds of thousands of dollars to monitor, but the cost of $3–4 per point, at, say, twelve points per motor/machine set is a bargain that is hard to pass up. In fact, one must be wary of the urge to monitor everything in sight, just because it is so inexpensive. Such an error will divert attention from the valid goals of the program.

Who Should Decide which Machines Should be Monitored?

The decision as to which machines should be monitored and the method of monitoring those machines (whether continuously or periodically) should not be made by the maintenance manager or the local vibration expert. This is a decision that must involve process, operating, and corporate marketing personnel as well. A logical way to proceed is as follows.

- An estimate of what quantities of what products will be made by the particular plant in the coming years must be prepared by the marketing personnel
- The above estimate will allow the process people (usually chemical or manufacturing engineers) to determine the likely usage of each machine to be used in the manufacturing process
- Knowing machine usage, operating personnel should be able to estimate a criticality number based on how necessary it will be for each machine to function without problems in order to meet the estimated need for product
- Maintenance personnel, knowing the relative reliability of each machine in the plant and the likely time required for a repair or replacement can then generate an approximate reliability index, which will indicate the order of priority for inclusion in the monitoring program for each machine
  Such a priority number might well be calculated as

$$\text{Priority} = \text{criticality} \times \text{reliability}$$

where criticality is assigned as a function of how necessary the machine is to the attainment of the production goals and reliability equals the inverse of the probability of failure of the machine when operated at the necessary operating conditions multiplied by the amount of time required to repair or replace the machine (i.e., time to replace/probability of failure).

How Often Should a Machine be Monitored?

The monitoring interval for a given machine depends on the operating history, design, and duty cycle of the machine. This interval is, at best, a guess that is highly tempered by the cost and difficulty of monitoring each location to be measured. Any monitoring interval short of continuous monitoring incurs the risk that an untrendable failure of some kind will destroy the machine between monitoring intervals.

As a first guess, intervals decided on the following basis might be reasonable:

- Machines with poor operating history, known design flaws, stop-start, or other severe operating conditions may have to be monitored daily or weekly.
- Most machines may only need to be monitored monthly or quarterly.
- Machines with proven long lives or minimal criticality may require monitoring only on an annual or semiannual basis.
- When a machine begins to exhibit problems, the monitoring interval should be reduced (sometimes hourly) until repairs can be made.

What Non-Vibration Parameters Should be Measured?

The most common thermodynamic properties measured are pressures, temperatures, flow rates, shaft speeds (usually obtained from the vibration measurement), and power consumption (measured in amperes or watts at the motor).

Bearing temperatures are often an excellent indicator of failing bearings. The method of obtaining this data varies from a thermometer held to a bearing housing by putty to a series of thermocouples implanted in the shoes of a Kingsbury-type thrust bearing. The degree of accuracy to be obtained is a function of the criticality of the measurement. Rolling element bearings, for instance, are less sensitive to changes in lube-oil temperature than sleeve bearings, which could go into shaft whirl or rub with improper oil viscosity. Some of the most recent vibration monitoring devices employ a pickup that can measure temperature and acceleration simultaneously.

Vibration measurements are generally fairly simple to obtain. As mentioned above, it is not a good idea to measure only unfiltered vibration. Also, as discussed in Chapter 3, careful consideration must be given to the choice of pickup to be used as well as the amplitude parameters to be measured.

The only parameters that should be measured are those that are nonintrusive to the manufacturing process. For instance, if one were going to monitor a process pump, the pressure differential across the pump would be obtained easily by adding pressure gauges near the suction and discharge of the pump. Obtaining the flow rate, however, might involve placing a flow meter or an orifice in the line causing unacceptable pressure drops in the process. The only reasonable indicator of insufficient flow rate through the pump in this case would be a drop in flow elsewhere in the system.

When deciding what parameters to use for monitoring a given machine, keep in mind that gathering too much data will result in a large data gathering effort and an unwieldy data trending system. If the maintenance manager and the vibration specialist are unsure as to what thermodynamic properties are most representative of the condition of a particular machine, the manufacturer should be questioned. Keep in mind, however, that with the possible exception of large steam or gas turbines, the manufacturer of rotating equipment is quite unlikely to be able to provide correct advice on the monitoring of vibration parameters. Most manufacturers lag equipment users by about a decade in competence in this area.

How are Baseline Criteria Chosen?

Thermodynamic operating criteria are easily obtained from either the equipment manufacturer, as just stated, the original purchase order specifications for the machine, or the plant process control personnel. Acceptable performance degradation before the machine must be scheduled for repair is a function of process requirements and the increased wear a machine will undergo due to such things as cavitation, stalled operation, and overheating.

Vibration baseline data is more difficult to obtain. There are three ways commonly used to obtain baseline vibration data: Many plants presently establish an almost arbitrary plant-wide level of about 0.1 in/sec peak. Although this method has the advantage that it is simple, it has the major disadvantage that it is too simple-minded to work. Not only does it ignore low-energy failure modes, such as in bearings or couplings, it ignores the entire nature of the individual machine.

A second, more acceptable method is to take filtered vibration readings at the measuring locations of each machine and assume that these levels are normal. A 6-dB increase (doubling) in the level in any frequency band will constitute an alarm state, and a 10-dB increase (tripling) will constitute a severe warning level. This method has the advantage that it can be put into effect very rapidly. Although it is true that some of the incipient failures will not be located because they existed at the beginning of the monitoring program, this will be offset by the statistical fact that most machines will be in good condition at the beginning and a great many machines will be monitored because of the simplicity of establishing the baselines.

Some further possibilities should be kept in mind with regard to this method. First, if someone has recently been working with a machine, or someone has a work station located near a machine, they should be asked to give their opinion as to the condition of that machine on inception of the program. Since even untrained people are aware of noises that bother them, as well as changes in these noises, their input is a valid reason to assume that a machine is either good or bad at the beginning of the program.

Also, remember that no baseline should be cast in iron. After a few monitoring periods have gone by, it is wise to reevaluate the baselines in light of the data. Computer programs exist to calculate the means and standard deviations of the gathered data to set statistically improved baselines.

A third method of baseline determination is to choose levels based upon a detailed narrow-band spectrum analysis of the machine. This will allow a skilled analyst to consider each of the mechanical components of the machine (which generates unique frequency components) separately. This is infinitely better than measuring an overall vibration level and hoping that it represents all the critical components (bearings, couplings, gears, rotors, etc.) of the machine. Each frequency peak must be examined separately and its amplitude level used to determine whether its forcing mechanism is occurring at acceptable levels of vibration. If not, the possibility of incipient machine failure must be considered and a detailed diagnosis of the problem carried out. The disadvantage of this method is the tremendous time it will take to perform a narrow-band spectrum analysis on each machine to be incorporated into the monitoring program. Many machines will have to wait weeks or months before inclusion in the program, and the building of their trended data will be delayed.

The methods of determining which levels of narrow-band spectrum analysis are acceptable can be found in Chapter 1. It is generally believed that, in the absence of other specific information, a doubling of an acceptable vibration level (6 dB) is an indicator of trouble and a tripling of the level (10 dB) is cause for much concern.

Under What Operating Conditions Should Readings be Taken?

If one is lucky, most of his process machinery will be constant-speed equipment operating at a fairly constant load. In this case, the conditions used for gathering both baseline and monitored data is obvious. Remember that even this simple mode of operation may show variations due to summer and winter operation. Also, since induction motors and engines can exhibit slight changes in speed, any computerized monitoring program must take account of changes in forcing frequencies via an order normalization scheme. Order tracking, as described in Chapter 2, is not usually required for constant-speed machinery.

If the equipment to be monitored does exhibit changes in speed or load due to changes in its operating requirements, the task becomes more difficult. One possibility is that there is a most probable range of operation that can be used for monitoring. If not, it is necessary to run tests to determine the variation in the measured parameters with speed and load. It has been the experience of the author that a great many kinds of fans and pumps do not exhibit much change in vibration level between 80 and 100% speed.

Where Should the Vibration Readings be Taken?

For rotating machines, vibration readings generally are taken at each bearing-housing system in the horizontal, vertical and axial direction. This information is useful because the vibration of the rotating equipment is transmitted through the bearings. (This is more pronounced with rolling element bearings than sleeve bearings.) As one gains experience in monitoring particular pieces of equipment, it becomes possible to reduce the number of data points on a given machine without significantly reducing the probability of finding an incipient fault in time to deal with it. Be sure to mark the exact positions where data are taken so that future readings may be taken in the same locations.

Vibration readings must be taken on a surface stiff enough that it is not affected by the pressure exerted by the probe or accelerometer. This phenomena is called mass loading by the transducer on the surface to be measured. Sheet metal, for example, will exhibit a completely different vibration mode when it is pressed. Such a reading would not be a valid indicator of how the machine is vibrating.

All vibration readings should be taken with the transducer perpendicular to the mounting surface. Vibration signals containing high frequencies must be taken with the accelerometer tightly screwed or glued to the surface, since hand pressure cannot hold it tightly enough to the surface to correctly follow such high frequencies. Magnetically mounting the accelerometer is preferable to a hand-held reading but not as good as a hard mount for high frequency responses. Most typical magnets can be relied on to yield accurate results up to approximately 3 kHz, if they are properly mounted. It is possible to purchase a super magnet, which, if mounted on a smooth surface, will have a flat response curve up to approximately 10 kHz.

It is often advisable to mount steel (or magnetic stainless) blocks at the transducer locations for future ease in screwing on or magnetically mounting the transducer. When periodically monitoring a machine that is not easily accessible, using a triaxial accelerometer rather than a single-axis device will reduce the number of times a technician must go from the measurement device to the transducer location (to change its orientation) from 4 times to 2 times per monitoring incident at each bearing.

There are some new accelerometer systems available that allow the mounting of several pickups at various inaccessible locations. The pickups are wired to a convenient node location for periodic monitoring. In some systems, the accelerometer can actually identify itself to the monitoring device.

What Kind of Vibration Transducer Should be Used?

Selection of a transducer is quite important to attain correct spectra on a vibrating machine. This topic is dealt with in detail in Chapter 3. A few comments will be repeated here:

If a proximeter is used, it is possible to look directly at the shafts of the machinery. It should be noted, however, that the frequency range for which valid information can be obtained is quite low, and a proximeter must be held rigidly with a clamping arrangement so that the proximeter does not vibrate due to anything in its environment.

Velocity pickups are popular because they have been around for a long time. However, they have some very basic problems. Since velocity pickups operate above their natural frequency (which typically is approximately 600 rpm), the sensitivity of the pickup drops quite radically below 600 rpm. Therefore, a velocity pickup should not be used to take measurements on a low-speed machine. In addition, the velocity pickup up tends to be sensitive to cross-axis affects. That is, if the pickup is mounted in the vertical direction, the output could be affected by extreme vibration in the horizontal or axial direction.

Accelerometers have the advantage of having adequate sensitivity over a wide range of frequencies. The low end of frequency response of a typical piezoelectric accelerometer is 1–3 Hz and the upper end tends to range between 5 and 20 kHz. For this reason, an accelerometer is often the preferred device to use.

If the same accelerometer is always used to monitor the same location, it is not necessary to use an accelerometer that is calibrated to a sensitivity of plus or minus five percent. A ten to twenty percent variation in sensitivity or a non-flat frequency response is of little consequence.

Certain problems can arise when accelerometers are used to measure low-frequency signals. These problems can be solved by integrating the signal (by analog or digital means) and reading out the vibration data in terms of velocity units such as in/sec (see Chapter 3 for more on this matter).

What Training is Required for a Successful Monitoring Program?

The technicians charged with carrying out the monitoring program can make it a tremendous success or a failure. The difference is in motivation. A technician who is required to simply carry vibration monitoring equipment from place to place and to press a prescribed sequence of buttons will soon realize that he can be replaced by an ape. He will begin to behave accordingly. The same technician, on the other hand, who is allowed to become part of the decision-making process concerning the health of the machines from which all plant employees make a living will soon acquire an elevated status among his peers. His performance will rise to meet their expectations.

With this in mind, the technician must be supplied with the best equipment available and a comprehensive training program consisting of a few days of vibration monitoring theory and months of encouragement and feedback. A job title denoting the importance of his work is also a help.

A Vibration Monitoring Application Matrix

Several fundamentally different approaches to dealing with the vibration monitoring aspects of a well-run predictive maintenance program have been covered in this chapter. A common mistake is to assume that these varied approaches are mutually exclusive. On the contrary, a large facility can optimize its predictive maintenance program by using a well thought out mix of approaches.

The proper technology for the predictive maintenance of a given set of machines can be decided upon based on the complexity, cost, and maintenance problems of a particular approach versus the cost and level of criticality of the machines being monitored.

A Monitoring Matrix
A Suggested Monitoring Technology Mix

Logistics / Criticality	SINGLE MACHINE	MULTIPLE MACHINES Close Proximity	MULTIPLE MACHINES Far Apart
Not Critical or Expensive	Ignore, replace at failure	Ignore, replace at failure	Ignore, replace at failure
Not Critical Fairly Expensive	Periodic Filtered Monitoring, or call a consultant as needed	Overall Continuous Monitoring, or Periodic Filtered Monitoring	Periodic Filtered Monitoring
Critical, Inexpensive, Spared or Easy to Rebuild	Periodic Filtered Monitoring, or call a consultant as needed	Overall Continuous Monitoring, or Periodic Filtered Monitoring	Overall Continuous Monitoring, or Periodic Filtered Monitoring
Critical, Inexpensive, Not Spared or Easy to Rebuild	Overall Continuous Monitoring, or Periodic Filtered Monitoring	Overall Continuous Monitoring with Filtered Data Collection	Overall Continuous Monitoring, or Periodic Filtered Monitoring
Critical, Expensive	Overall Continuous Monitoring, supplement with Periodic Filtered Monitoring, or add on-line Filtered Data Collection	Overall Continuous Monitoring with Filtered Data Collection	Overall Continuous Monitoring with Filtered Data Collection

Often, a plant will have machines requiring different levels of predictive maintenance. One can mix the various aspects of vibration monitoring technologies to best fit the application at hand. Periodic monitoring, for instance, can be carried out using an overall meter with some small success. For far better success, an FFT-based data gathering device, combined with a computer-based monitoring program can be used. Likewise, an overall vibration-based continuous monitoring system is commonly used with success to prevent the destruction of a costly machine. This system can be enhanced by adding FFT-based filtered data and monitoring software to enhance the predictive nature of the program.

The cost of enhancing a simple system may require only a small incremental cost or become quite expensive and unwieldy, depending on the application. The Monitoring Matrix presented here is designed to show the relationship of various technologies as a function of the application at hand. A discussion of the logic behind the Matrix, as well as a description of the items shown in it, follow.

Logistics

Although most handheld portable vibration monitoring devices lend themselves to being carried large distances from machine to machine, continuous monitoring requires a great deal of expense and effort to set up in widely diverse areas of a plant. Therefore, a distinction must be made between machines which are physically far apart and machines which are grouped together, making the cabling necessary for a continuous monitoring system more economical on a per machine basis.

One of the major capital costs of setting up a continuous monitoring system lies in the cost of running cables. In the past, it has been necessary to run a cable from each transducer location to a control room for continuous monitoring. If it was desired to enhance the overall readings available at panel meters in the control room, one or more scanners would be connected from a number of panel meters to a computer-controlled FFT-based vibration analysis system. Although this sort of plan has the obvious advantage of adding periodic narrow band condition checks (predictive, periodic) to the simple overall alarm checks (nonpredictive, continuous), the cost of cabling remained high. This cost had hampered the use and further development of continuous monitoring systems, except where absolutely necessary.

A recent addition to the art of machinery monitoring has been made in the area of cabling. With the introduction of a device patented by the Vibra-Metrics Corporation of Hamden, CT, it has become a simple matter to connect a very large number of accelerometers to a computer-controlled FFT-based vibration analysis device in the control room with a minimum of cabling. The device, called a Sensor Highway, uses a single cable network to connect all accelerometers. Each accelerometer has its own digital address (See Figure 6-3). The control room analysis device can access each accelerometer in the

A Sensor Highway wiring layout.

B   Sensor Highway controller.

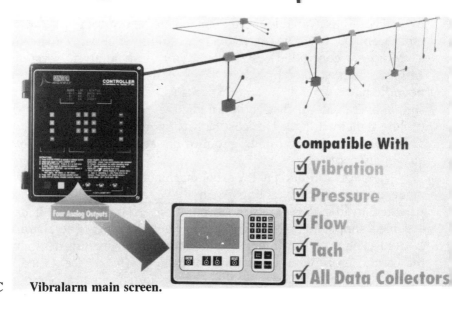

C   Vibralarm main screen.

Figure 6.3.   Sensor Highway.

network in any sequence desired for FFT analysis. Although this system is not continuous in the classic sense—only one accelerometer is accessed at a time—it simulates continuous monitoring quite closely while offering the advantages of FFT-based predictive maintenance. One might consider this system as a predictive periodic monitoring system with a very short access period.

Criticality

There are three aspects of criticality which must be considered when entering the Matrix. These are component replacement cost, the cost of lost production on a machine, and, probably most important, the potential for damage to the rest of the process machine.

The question of how costly a machine is usually depends on the size of the capital budget of the facility. The decision to allow a $30,000 pump which is physically isolated from the rest of the process equipment to run to destruction in order to meet a critical demand period is not uncommon in large plants. The $30,000 needed to replace the above-mentioned pump could severely damage the Profit and Loss Statement of a small company.

Lost production cost is a function of the value of the product in the marketplace and the effect a given machine has on the volume of production of the product. An analysis of how much of the total production will be lost due to the outage of a particular machine must be estimated.

A failing roll bearing in a process machine such as a paper machine could easily destroy the entire machine in seconds. Signs of incipient failure in such a component should be treated far more cautiously than for the isolated pump mentioned above.

In some special cases, the degree of criticality cannot be estimated in dollars. The failure of safety equipment in a nuclear plant can cause months of costly investigation and analysis of the problem, as well as a considerable increase in the probability of danger to plant personnel and local residents. An air handler which cools the office of the company president is far less critical to corporate shareholders, but very critical to the maintenance manager who relies on the president for his next salary increase.

A machine becomes less critical if it is spared. This is why it is common for a power plant, for instance, to employ several smaller water circulating pumps rather than one large one sized for full capacity. A dryer in a paper machine cannot be spared and is thus quite critical to the entire process.

Ignore, Replace at Failure

Some machines are so inexpensive, easy to replace, or superfluous to the fundamental operation of the plant, that it is not reasonable to monitor them at all. It is far more efficient to stock a unit in anticipation of the day of failure. The effort of monitoring such a unit simply diverts attention from the monitoring of critical machinery.

Overall Continuous Monitoring

A continuous monitoring system is designed to look at every transducer continually. The message one receives from such a system, therefore, is one of a warning of a problem of some significant level, followed by a shutdown if the problem reaches a level wherein the machine is about to undergo severe damage if allowed to run for even a few minutes longer. The opportunity to undergo a planned shutdown for maintenance when convenient seldom exists in the typical continuous monitoring system.

Because a continuous monitoring system is comprised of a great many permanently mounted transducers and long distance hard wiring, the cost of installation and maintenance of the system is rather high. This fact, along with the system's inability to provide a sufficiently long early warning of incipient failures, results in the conclusion that continuous monitoring is best employed only in cases where it can be implemented relatively inexpensively for the protection of costly machinery.

Periodic Filtered Monitoring

Periodic monitoring involves the gathering of vibration data at selected locations, on selected machines, at a regular time interval. The data gathering device may be a simple-minded overall meter or a sophisticated FFT-based device. For the reasons discussed in describing the difference between overall and filtered vibration readings above, it is hard to imagine a case where periodic overall machine monitoring is justified. The cost of an FFT-based monitoring device is small enough that the use of an overall meter should not be considered.

The major cost of a periodic monitoring program is the salary and benefits of the person carrying the handheld data-gathering device from place to place. This is often less expensive than installing a great many transducers on a machine and running cables to some distant control room. One of the major advantages of periodic monitoring lies in having someone physically look at the machine on a regular basis.

If a periodic monitoring program using in-house personnel is to be successful, it is imperative that those charged with the duty of data gathering be properly trained in both the operation of the data-gathering equipment and the general behavior of machinery. The technician should be able to download routes into the monitoring box and upload data into the computer. If the technician is not encouraged to try his hand at diagnosing problems indicated by the computer, he will soon lose interest in the project. The quality of the program will then suffer.

Many plants have found that the ability to keep trained people on hand and available to perform periodic monitoring on a regular basis is too limited. A viable solution to this problem is to contract with a firm that specializes in periodic vibration monitoring. An additional advantage to this course of action is that, when a problem is located, additional minds can be applied to it.

Overall Continuous Monitoring With Filtered Data Collection

This represents an enhancement of the normal overall continuous monitoring system. The same alarm and shutdown criteria are employed on the unfiltered transducer signals, but, in addition, a scanner automatically switches each signal to an FFT-based analyzer and computer. In this way, the machine so monitored can be continually protected against a rapidly developing failure while advanced notice of an incipient failure can be had. Systems such as this are sometimes found in large refineries and can cost from one hundred thousand to several million dollars for instrumentation, installation, and programming.

A less costly method of obtaining both immediate protection and diagnostic capability is to install an overall continuous monitoring system while adding the machine to a periodic monitoring machine route.

Call A Consultant

Single machines often do not justify inclusion in a regular monitoring program due to their inherent reliability, lack of criticality, or low replacement cost. It is sometimes advisable to simply wait until a problem with a machine is suspected due to unusual noise or a drop in performance and call a consultant to diagnose the problem. Since there would be no previous vibration history on the machine, the consultant should be expected to have very sophisticated equipment, including the capability of dual channel analysis as well as a good mechanical engineering background.

What Next?

Having made the important decisions discussed above, it is now time to give consideration to those topics concerning equipment which must be purchased. Chapter 7 will discuss the kinds of monitoring, trending, and diagnostic hardware available. Some criteria for the selection or writing of software for computer-based monitoring and trending programs will also be given.

Tomorrow is too late to begin planning a monitoring program. Since the proper equipment, training, and historical data is essential to conducting a monitoring program, to the program's being successful, and to obtaining the highest benefit/cost ratio, today is the time to start planning for a monitoring program. Yesterday was not soon enough.

The initial decisions discussed above, as well as selection of the proper equipment, should start immediately. Personnel training should start as soon as possible. A poor economic climate or low product demand is an ideal time to begin. It is in these hard times that a major machine failure will put the biggest drain on the financial resources of the corporation. This is the time to be efficient. People are more likely to be available for training during slack production periods so that they will be ready when product demand increases and machinery is run at full capacity again.

CHAPTER SEVEN

Hardware and Software

Introduction

The quality of the data gathered for monitoring, trending, or diagnostics, and how it will be used, will obviously effect the success of the predictive maintenance program. In this chapter, the equipment and software necessary to do a proper job in each of the three phases of rotating machinery analysis is discussed. Transducers, however, will not be discussed here, because they were dealt with in detail in Chapter 3. Although specific products may be mentioned occasionally in this chapter, the intention is to make the reader aware of the kinds of things available. No endorsement of a particular product is intended.

The electronics and software industries are different than the heavy machinery industries in two ways, which surprised the author when leaving a job with a heavy-equipment manufacturer some years ago in favor of an electronics instrumentation manufacturer. It would be wise for the machinery people among us to learn these lessons now:

Because of the frequency of introduction of new electronic chips, the price of equipment using older technology begins to rise after a few years. This means that an old, reliable design can no longer compete with a more recently designed instrument in terms of cost, weight, or (sometimes but not always) capability. This forces instrument manufacturers to introduce new equipment every 3–5 years instead of every 10 or 20 years for a pump, gearbox, or turbine. Any specifics about particular pieces of equipment included in this chapter, therefore, are on their way toward being obsolete as they are written. Obviously, this is even more true of software.

Usually, serial number one of a pump or turbine model is very close to being correctly designed. Any minor problems are fixed via retrofit early in the product life cycle of the machine. Electronic devices and software are not like that. It is not uncommon, in fact, for an instrument with revision E or F improvements and a year in the marketplace to still have significant problems. For this reason, unless a new piece of equipment is substantially based on an old design, or can be easily updated by loading in a new software revision, one should resist the temptation to buy it for a year or two. Let someone else have the problems.

If the above advice seems excessively cynical, ignore it at risk of your vibration program. After spending more then $20,000 on a piece of equipment or a software package, one can be sure of having to live with it for a very long time before the purchase of new equipment is approved. Also, a malfunctioning instrument does not help the buyer build credibility with his supervisors.

The Vibration-Monitoring Device

Most vibration-monitoring programs begin with a simple overall velocity meter. While better than nothing at all, the success rate of the monitoring program can be improved greatly by using filtered measuring devices such as octave or 1/3-octave meters, tunable filter analyzers (which tend to be too tedious to operate), or a spectrum analyzer. The most recent devices, which will be discussed in detail, are handheld FFT-based boxes with large nonvolatile memories.

Since the reader must be aware by now that unfiltered velocity meters are not a good way to gather data for a periodic monitoring program, these devices will not be dealt with at all. Instead, the devices covered will begin with the octave band meter as the simplest acceptable meter for monitoring and proceed to more complex devices.

Octave and 1/3-Octave Band Meters

Octave band data is probably the widest set of filters one would want to use for general machine work. Since an octave is a doubling in frequency (see Chapter 1 for a list of standard octave filter sets), it is likely that a forcing function at 1 × rpm, for instance, will be separated from a 2 × rpm misalignment signal, or a 4 × rpm blade frequency as well as a 58 × rpm gear mesh frequency as long as the shaft speed is low enough to cause all of these forcing frequencies to fall in relatively low center frequency filters. For motor driven machines, this is the case since the highest possible 60-Hz synchronous motor speed is only 3,600 rpm. One would have trouble monitoring a device mounted on a helicopter operating on 400-Hz line frequency power.

There are ten standard octave bands from 10 Hz to 10 kHz, requiring the trending of only 10 filtered levels plus one overall level per reading. This means that it is feasible to record the data with a pencil, archive it with a file cabinet, and trend it with a secretary.

Even though the trending of 11 numbers (including overall level) per transducer location can be done in the smallest of computers, when it becomes obvious to management that a computer can do the job more efficiently than a secretary, a problem with octave meters begins to surface. No octave band meter yet manufactured includes the capability of digitizing the 11 numbers and storing them in internal memory by transducer location number for transmission to a computer. Octave-band computer-based monitoring requires a human to type the data into the computer. Such a method is expensive, slow, and fraught with errors.

Obviously, a 1/3-octave (23%-bandwidth) filter set will be more definitive than an octave filter set due to increased resolution. It will be possible to separate more forcing frequencies, reducing the probability of masking a low-energy phenomenon such as a ball bearing frequency from a blade frequency 50 Hz away. The 1/3-octave filter set will require the trending of 30 filtered levels, plus an overall per reading from 10 Hz to 10 kHz. The same comments regarding recording, archiving, and trending the data apply as with octave meters.

Tunable Filter Meters

Some manufacturers build tunable filter vibration meters with 10 or 20% filter bandwidths. Although these filters have adequate resolution for a large number of machines, they tend to be very tedious to use.

The type of tunable filter instrument that requires the technician to manually tune a filter while looking for peaks on an amplitude meter is totally unmanageable as a monitoring device because of the time-consuming nature of finding and recording an undetermined number of localized peaks by hand. The problem of entering a set of amplitudes and corresponding frequencies (which will almost certainly be somewhat different each time a machine is examined) is so large that any thought of computerizing the trending function for this kind of data should be abandoned.

Some tunable meter devices incorporate a plotting device, which yields a hard-copy amplitude versus frequency spectra for each transducer location. These plots are adequate for comparison of one spectra to another by holding two of them to the light. Many people have fallen into the trap of mounting dozens and dozens of these charts into books only to find that data retrieval and trending can take hours and hours by this method.

FFT Spectrum Analyzers for Monitoring

Some plants use an octave or 1/3-octave analyzer to locate machine problems and a spectrum analyzer (due to its more powerful diagnostic ability) to examine the problem for decision-making purposes. Some plants use a spectrum analyzer for both monitoring and diagnostics. It is often true that the most economic course of action is to monitor with a simple octave or 1/3-octave meter and use an outside consultant for complex diagnostics.

Although the FFT spectrum analyzers are the most powerful of analysis devices and are capable of interfacing to a computer for trending, there are some physical problems that make them cumbersome for periodic monitoring. Note here that we are talking about full capability spectrum analyzers, not hand-held FFT boxes. Over time, these problems tend to become larger and larger sources of irritation to the data-gathering technician.

Most spectrum analyzers use 50-60 Hz line power. This means that someone wishing to take, say, 200 readings on various pieces of equipment in a day, is likely to have problems finding convenient power outlets near all of the machinery of interest. If the analyzer does not have the capability of storing several panel setups in nonvolatile memory, the analyzer will have to be properly reconfigured each time the unit is moved to a new power source. To date, only the most limited capability analyzers weigh less than 35 lb.

In most plants, the amount of traveling from one machine to the next requires that a device weighing more than 10 lb be mounted on a cart, causing other logistic problems.

Analyzers do not tend to be certified explosion proof. In many areas of a process plant, for instance, either a "hot work permit" must be obtained or cables of several hundred feet between the analyzer and transducer must be run. Furthermore, many analyzers (especially those with wire wrap boards rather than printed circuit boards) cannot be subjected to the hostile environments common in modern industry (another reason for long cables).

Storage of the data gathered by the analyzer is a problem. Some of the newer analyzers have optional bubble memories or battery backed CMOS memories (which one makes no difference to the user). Unless the particular analyzer has enough memory to save as much data as a technician can gather in a day, peripheral equipment such as tape or disk drives or plotters must be carried along with the analyzer. Further comments on data storage devices will be made later in this chapter.

Even if a computer is used to trend and archive the monitored data, the fact that 400 lines of data are gathered for each transducer location in the monitoring program can become a problem in terms of data storage and computational time. More about this problem will be discussed below.

Handheld FFT Devices

Starting in 1982, research was begun on handheld data gathering devices for periodic monitoring. The initial work was done at Exxon in terms of gathering thermodynamic data relating to machine condition and at the now defunct Nicolet Scientific in terms of filtered vibration instrumentation. Because of recent developments in low-power A/D converters, an FFT chip, and the rapid drop in CMOS memory costs, several devices now exist that fit the original Exxon/Nicolet notion of a filtered handheld monitoring device.

In order to give the reader a feel for the way in which these handheld FFT devices work, a typical day in the life of a monitoring technician using one will be described:

> 7:00 am: The technician connects the FFT box to the computer and instructs it to load the FFT box with information for route 2.
>
> 7:01 am: The FFT box is loaded with the following data: (1) The machine numbers and transducer locations for all of the machines on route 2; (2) acceptance criteria for each of the transducer locations on the route—each criteria can be composed of up to 400-line vibration levels and a number of thermodynamic inputs such as pressures, temperatures, power consumption, etc.; and (3) panel settings for each transducer loca-

tion—such items as number of averages, frequency range of interest, number of lines in the FFT, level of integration (acceleration, velocity, or displacement) are entered to save the technician the trouble of setting up the instrument on site.

7:30 am: The technician arrives at machine number 123456 and enters that number into the FFT box via the keyboard or a bar code scanner. The box tells the technician via an LCD display (or vice-versa) that transducer location ABC is to be taken and asks if the transducer is properly mounted. The technician gives an affirmative response and the FFT box takes in the proper vibration (and, possibly bearing temperature) data.

7:31 am: The FFT box stores the data just taken in for later transmission to the host computer. It also compares the data to the baseline criteria and informs the technician of the results. On some design boxes, the technician has the option of going into a scratch-pad mode to gather additional data if a problem is found in an attempt to solve the problem on the spot.

7:32 am: The FFT box requests the technician to read several temperature and pressure gauges in the vicinity of machine number 123456 and stores the data after comparing it to baseline values. Note that, at any time, the technician can interrogate the FFT box to find out what machines remain to be done on route 2.

3:00 pm: After having taken data at 150 different transducer locations during the day, the intrepid technician returns to the host computer and commands it to trend and archive all of the data entered into the FFT box during the day. Depending on the software in use, various condition exception reports will be issued to the maintenance department and management concerning the machinery on route 2.

As can be seen, the FFT box is, by far, the most convenient and foolproof monitoring device available. Be forewarned, however; some of the manufacturers of these boxes claim that the purchaser of their boxes need not purchase a full-sized spectrum analyzer or hire a consultant to fully analyze a problem that has been located during the monitoring program. Except in very simple cases this is not true. The small FFT boxes lack the capability to do such things as zoom, order track, synchronous time average, or even provide such niceties as harmonic and sideband cursors for a proper analysis. Do not confuse a monitoring box with a diagnostic box.

Some of the capabilities to look for in an FFT monitoring box are listed below. Not all of these are necessary for every application.

Antialiasing Filters Remember that a handheld FFT box is subject to the same physical laws as the full-sized spectrum analyzer describer in Chapter 2. There must be adequate antialiasing filters, sampling rates, and windowing to get good data. Some manufacturers of handheld boxes seem unaware of this fact. It is necessary for the potential user to evaluate a particular product with a specification sheet in one hand and Chapter 2 in the other.

Dynamic Range Many handheld monitoring devices have a dynamic range of only 48 dB. Although this may be enough in many cases, a 72-dB unit will have a better chance of finding a low-energy bearing fault in the presence of a high-energy gear mesh signal. Devices with less than 48 dB should be regarded with suspicion.

Autorange It is not reasonable to expect a technician who may be making over a hundred measurements a day to take time to correctly set the attenuators that protect the antialiasing filters before each reading. It is therefore necessary for the FFT box to automatically set them at the start of each reading. Autoranging in 10-dB steps is adequate.

Frequency Ranges An FFT device has a number of equally spaced filters; a problem therefore arises that is sometimes ignored. If a wide frequency range is required to see a high-frequency forcing mechanism, such as a turbine blade frequency or a gear mesh, the resolution of each of the filters might be too broad to separate a running speed signal from twice running speed. For example, consider an 900-rpm machine with 400 teeth on one of its gears (such as might be found in a ship drive) such that the running speed is 15 Hz, bearing oil whirl is approximately 7 Hz, twice running is 30 Hz, and mesh frequency is 6,000 Hz. Therefore, the frequency range of the box must be 0–10 kHz with a resulting bandwidth of 10,000/400 or 25 Hz. As can be seen, there is not quite enough resolution to follow the condition of the low-speed forcing phenomena of the machine in the same spectrum as is required to see gearbox condition. All of these important forcing phenomena can only be observed if the FFT box gathers two sets of data, one at, say, 0–500 Hz and another at 0–10 kHz. Another possible scheme for handling this problem will be discussed in the Software section. A few of the newest portable FFT boxes meet this problem by providing a 1600 line transform. This slows the monitoring by a factor of four.

Averaging The FFT box should be capable of performing a number of summation averages before comparing or storing the data.

On-Site Comparison It is not enough for the box to simply store data for transmittal to the host computer. It should also be capable of doing an on-site comparison of the data with acceptance criteria for the particular transducer location. This alerts the technician to make any extraneous observations that may be helpful in the resolution of an indicated problem. Immediate feedback of this type also serves as a motivator for the technician. In plants too small to justify a technician spending adequate time learning vibration analysis, this capability is unnecessary.

Data Storage The box must have adequate data storage to hold as many spectra and acceptance criteria as a technician can visit transducer locations in at least half a day. As a corollary to this, the batteries must last at least 4 hr with a memory back-up battery good for several months.

Ease Of Operation The box must communicate with the technician in English. The display must be alpha-numeric and on an LCD display capable of being seen in sunlight. An elaborate display of a spectrum or a sketch of the machine may not be worth the additional bulk or expense. The box should not weigh more then 5 lb and should be small enough to safely carry about the plant on a shoulder strap. The box must operate in hot, cold, wet, and possibly explosive atmospheres. Furthermore, it should contain circuitry to turn itself off if the battery gets too low to acquire good data.

Hardware and Software 159

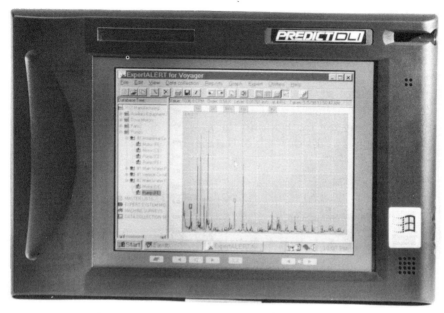

Figure 7.1. The Predict OLI Model OCX is a Data Collector/Analyzer using Windows 98, 16 mb of RAM, and a 2 gig hard drive for on-board monitoring.

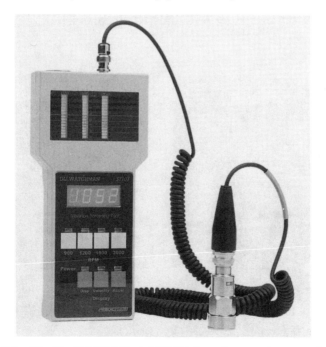

Figure 7.2. The Predict OLI Model ST-101 is an inexpensive screening tool which can be used to locate simple problems.

Interfacing The box should contain an ICP power supply for an accelerometer as well as an analog input for other kinds of sensors. The output should be either RS-232 or IEEE-488 for communication to computers, modems, or printers.

Special Sampling Techniques The special sampling techniques described in Chapter 2, zoom, order tracking, and synchronous time averaging may not be useful in the monitoring box. These functions are usually left to the domain of the diagnostician.

Software for Trending

Although it is possible to write one's own software for the monitoring, trending, and archiving of the data gathered in one of the handheld monitoring boxes described above, there is not much sense in reinventing the wheel. The high cost of specialized technical software may seem staggering at first, but if one were to calculate the cost in man-hours to write and debug one's own program, the expense may begin to seem reasonable. In this section, some worthwhile features of monitoring software will be discussed that may or may not be available in the particular programs being considered by monitoring personnel.

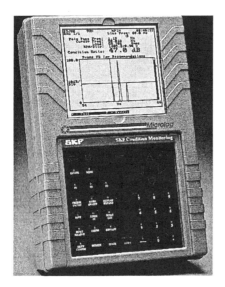

Figure 7.3A. The SKF Microlog Condition Monitor

Figure 7.3B. The CSI 2120 Machinery Analyzer

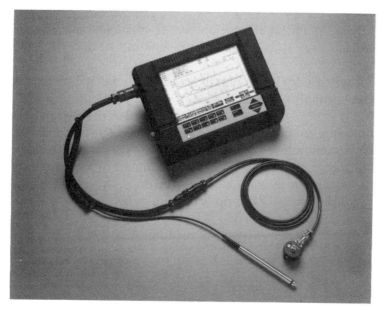

Figure 7.4. The Predict DCI Model DC-7B simultaneously measures three channels of vibration data and can double as a field FFT Analyzer.

Most periodic monitoring programs include trending capability. Trended data enables an intelligent decision to be made as to how much longer a failing machine may be operated safely. This ability makes trending very worthwhile. There are several possible formats for displaying trended data.

1. Overlays of data on amplitude versus property axes (the term property is used because the horizontal axis need not only be frequency; the plot can include level versus various pressures or temperatures, for example)

2. Plots of the amplitude of a particular property over time (this is quite useful because it allows the user to extrapolate the data to estimate time to failure)

3. Plots of the rate of change of the amplitude of a property over time (this plot is more sensitive to small changes in level than the plot just described)

4. A statistical evaluation of the data and an estimate of time to failure based on regression theory

Since trending relies heavily on historical data, the ability to store and retrieve data rapidly is crucial. This may be done with the use of plots, tape-recorded data, or a large number of computer floppy disks. In almost all cases, however, it is soon realized that the most efficient way to handle the large amounts of data necessary is via a hard disk.

There is very little point in amassing so much data on a piece of machinery that retrieval becomes a major problem. In some programs, after the baseline data has been verified over a period of time, the monitoring function continues but no new data is added to the trending system unless it is different from the previous information.

Another method of limiting the amount of stored data is to save new data but to average away old readings. One possible scheme for doing this, by way of example, is as follows:

- Take four weekly sets of reading
- At the start of week 5, calculate a monthly average, discard the first week's data, and gather data for week 5
- At the start of each week, update the monthly average; discard the earliest remaining weekly reading, and add a new most recent reading
- At the beginning of a new year, use the same process to remove monthly averages

Note that, after the data base is fully established, there will always be four weekly sets of data, twelve monthly averages, and a number of annual averages. This amount of data does not make excessive demands on one's computer memory.

The most powerful vibration measuring tool, the FFT device, will yield 400 pieces of vibration information per reading. To avoid archiving very large amounts of data, many plants employing spectrum analysis for monitoring use a desktop computer to do a process called *peak picking*. A peak-picking program causes the computer to look at the 400 lines of information in a spectra and store only the significant peaks for trending.

There is still another way of reducing the amount of data that must be archived and manipulated for monitoring and trending. This is to use variable-width filtering. As was mentioned with regard to the FFT monitoring device, even 400-line data might not be narrow enough to distinguish between various forcing frequencies. One possibility is for the vibration analyst, knowing the characteristics of the individual machine to be monitored, to create a set of filter requirements for that one machine. The software program would then instruct the handheld FFT box to take data in certain frequency ranges, which the host computer would then selectively add to simulate the ideal set. This could easily result in a set of 40 or fewer amplitudes to be stored rather than 400. If the handheld FFT box has adequate program and programming capability, in fact, this method could be implemented "on board" with variable bandwidth comparisons and more efficient memory utilization in the handheld FFT box.

In addition to trending vibration data, it is beneficial to trend thermodynamic data describing the condition of the machine. It is sometimes more informative to store calculated data than measured values. Thus, for a pump, for instance, pressure differential may be more valuable than P_1 and P_2, the inlet and discharge pressures. Efficiency may be more useful than horsepower. These decisions should be made at the outset of the monitoring program for each machine.

If certain data are trended, plant operation can be evaluated. For instance, often, when spared equipment is available, two or more identical units are installed in such a way, that, for stop-start or peak capacity operation schemes, the machines take turns acting as the lead unit. Suppose, for instance, that two pumps were connected in parallel with a stop-start operation. The first time the system is energized, pump A turns on. Pump B is off. The second time the system is needed, only pump B turns on while pump A rests. Both pumps operate simultaneously only when unusually large flow rates are required.

Suppose, further, that the monitoring technician goes to the pumping system weekly to gather data for trending. If the automatic alternating system fails to switch the lead pump function between the two units, the technician may well measure data on one of the pumps for weeks before realizing that there is a problem. During this time, one pump is wearing itself out, while the other pump is standing idle without the protection of being properly decommissioned. A good trending system would be able to pick out certain operational problems such as this by keeping track of which units were running or not running each time the data for that route was scheduled to be taken.

Another item that a complete trending system should check is exact replication of data (either vibrational or thermodynamic) at each data-gathering incident. Nature does not usually permit exact repetition of data unless a gauge or transducer is broken.

The trending software should also allow for operator entered notes such as the fact that a given piece of instrumentation on a machine has still not been repaired or that certain work was done to a particular machine. These notes can serve to alert management to things that aren't getting done or point out the need to establish new baselines for rebuilt equipment.

Spectrum Analyzers

When a problem in the operation of a machine is located, the proper thing to do is to diagnose the problem completely. This will allow for the proper ordering of new components and scheduling of equipment and personnel to do the repair with minimum downtime. A proper diagnosis is likely to reduce the occurrence of repeat failures.

The only device powerful enough to do a proper vibration diagnosis on most heavy machinery is a spectrum analyzer. The following are desirable characteristics to have in an FFT spectrum analyzer to be used for diagnostics:

- 400-line resolution
- At least a 0–10 kHz frequency-range capability, 0–50 kHz is preferred
- At least 70 dB true dynamic range
- Adequate antialiasing filters (see Chapter 2) to insure noncontamination of the correct signal anywhere in the spectra, over the entire dynamic range (Note: For a 400-line device with a sampling rate of 2.56, antialiasing filters with a rolloff of 140 dB/octave are required)
- Sum and exponential averaging (on some units, sum is obtained via the exponential averaging function)
- Zoom translation
- Peak hold
- Order-tracking capability
- Synchronous time averaging
- Transient capture
- Two-channel (transfer function, coherence) capability with at least 200-line resolution
- "Live" memory of a one or more other spectra for comparison
- Harmonic and sideband cursors
- IEEE-488 interface capability to computers and interface capability to plotters, printers, and other data storage devices

Before purchasing a spectrum analyzer, read Chapter 2 again. Also call some knowledgeable people who do not work for spectrum-analyzer manufacturers for their opinion on a given unit, as there is no substitution for experience in making such a complex and costly investment.

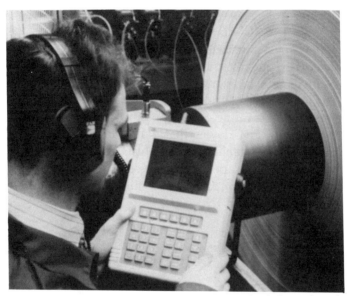

Figure 7.5. The Diagnostic Instruments Di-2200 dual channel analyzer.

Recording Devices

Besides storing data as digital information on computer disks, several older data storage schemes are still being used. Aside from plotters, the most common devices—tape recorders—are the most frequently used. Since there are frequently more errors made in the choice of which (if any) tape recorder should be bought, they will now be discussed in general terms.

Analog Tape Recorders

Two kinds of tape recorders are commonly used for vibration recording. Analog tape recorders are connected directly to the transducer signal. It is played back for either monitoring or analysis at a later time. Digital tape recorders are connected to the output of an FFT spectrum analyzer and are used to save the data seen on the display screen of the analyzer.

166 Chapter Seven

One school of thought dictated that, to avoid ruining one's spectrum analyzer by carrying it into a paper mill or chemical plant, an analog tape recorder should be used to gather data and bring it back to the office. The weak link in this system was the quality of the tape recorder and the ability of the operator to gather correct data.

Since, in this system, the data that is analyzed is only as good as the tape recorder, the recorder must be of the highest quality. It should have at least 60-dB signal-to-noise ratio, have very accurate tape transport speeds (usually reel to reel), have an FM tape head for recording low-frequency data and a direct tape head for high-frequency (over 500 Hz) work.

One of the major drawbacks to analog tape recording is lack of sufficient dynamic range. Some (possibly important) peaks can be seen using a 72-dB FFT device that would be lost to the technician who chose to tape the signal before analysis. It is also more difficult to operate and calibrate a tape recorder system. Many monitoring programs have ceased to function because the technicians did not properly use the tape recorder.

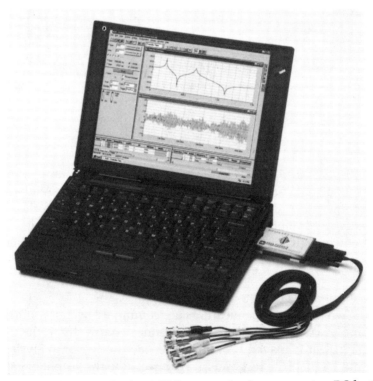

Figure 7.6. The Data Physics ACE Spectrum Analyzer converts a PC laptop into a dual channel analysis instrument.

Hardware and Software 167

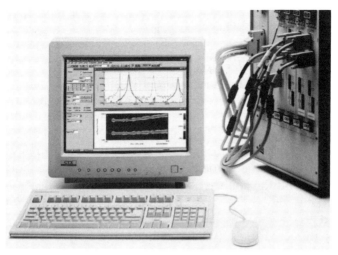

Figure 7.7. The Data Physics Corp Model 620 has up to 120 input channels and 90 dB Dynamic Range.

Figure 7.8. Larson Model 2900 Spectrum Analyzer

Because of advanced design techniques and modern electronic components, reliable high-quality FFT vibration analyzers have been introduced that cost considerably less than analog tape recorders but do not distort the data. The old notion of using a tape recorder to protect the spectrum analyzer from abuse is no longer valid. The only remaining advantages of an analog tape recorder are that one can change analysis ranges or zoom on the data months after it was taken and that some tape recorders can record 14, 28, or more channels of data simultaneously for later analysis.

Digital Tape Recorders

Because a digital tape recorder does nothing more than preserve the digital data representing the information and annotation displayed on the screen of a spectrum analyzer or other device with a digital output, calibration, dynamic range, and other factors influencing the accuracy of the stored transducer data are not relevant. The information retained on tape is always as good as the technician, the transducers, and the analyzer.

Many digital tape recorders use economical cassette tapes or disks and hold a very large number of spectra. Most of these units can be computer controlled, which can greatly simplify the trending process.

The main disadvantage to a digital tape recorder is that it is not possible to look at the data in ways different from the way in which it was stored. Any procedure that would require resampling on a spectrum analyzer (such as a change in frequency range, zoom, etc.) would not be possible with existing digital tape recorded data.

Summary

The purpose of this chapter was to provide the reader with a large list of considerations to keep in mind when making purchase decisions concerning equipment and software for periodic monitoring programs. Armed with this list, and any other concerns the potential purchaser might have, it is the obligation of the purchaser to study the equipment specifications from many potential suppliers. Do not be afraid to be hard on the sales people. If they cannot help you now, they probably will not be able to help you when you have a problem. Now is the time to find out which salespeople will be able to support you and which will never want to see you again after their commission has been paid. Although salespeople come and go, some companies have a history of standing behind their products and try to be helpful after the sale has been made. If possible, get a number of references from the prospective vendors and take the time to talk to each one personally.

Remember that the desired end result of your purchase is a complete system of equipment and software. It must perform well with your well-designed monitoring plan (Chapter 6) and your well-trained technicians. It is your reputation that is on the line.

CHAPTER EIGHT

Advanced Analyzer Functions

Introduction

In the following pages, we will review the previously covered functions commonly found in an FFT spectrum analyzer in further detail. Less commonly used functions, such as power spectral density, special time windows, impulse response, and inverse transfer function will then be covered. The reader will find the definitions and explanations in this chapter to be useful when a test has already been run and the results are unintelligible. If this chapter is reread at that time, it should be possible to redesign the test to get more meaningful results. Other functions, which are available on some spectrum analyzers but are not really frequency domain properties, will be covered in Chapter 9.

Complex Numbers

To gain a more in-depth understanding of the functions available in FFT spectrum analyzers, it is necessary to have at least a vague notion of the meaning of complex numbers. Complex numbers came about because no one knew what else to do about finding the square root of minus one. Since $-1 \times -1 = +1$, it follows that $\sqrt{-1}$ is not -1. Furthermore, since $+1 \times +1 = +1$, $+1$ is not the answer either. What to do? Mathematicians solve the problem by assigning a value of i to $\sqrt{-1}$. Because electrical engineers use i as the symbol for current, they use j as their symbol for $\sqrt{-1}$. Taking the notation of the electrical engineers

$$j \times j = -1$$

170 Chapter Eight

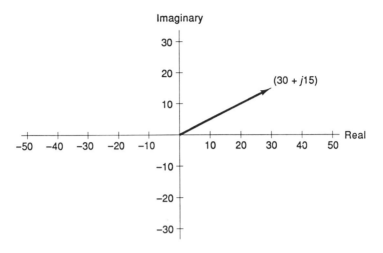

Figure 8.1. Plotting the complex number 30 + j 15 on the real-imaginary plane.

The name given to *j* is the imaginary part of a number. It may well seem ridiculous to invent an imaginary set of numbers (as opposed to the real set we are used to), but keep in mind that the only reason for worrying about *j* is that it keeps showing up in equations. If we learn how to manipulate it, we needn't try to understand the mathematics behind it.

A complex number has a real part and an imaginary part; for example 30 + *j* 15 can be plotted (Figure 8. 1), if we recognize that a complex number can be plotted as a location on the real-imaginary plane. As soon as this is recognized, it becomes apparent that complex numbers can be handled in exactly the same way as are vector mechanics. The general form of a complex number may be written as *X* + *jY*. We can then write the following:

$$R = X + jY$$

The real part of *R* is *X* and the imaginary part of *R* is *Y*. The magnitude of *R* is

$$|R| = (X^2 + Y^2)^{1/2}$$

The phase of *R* is

$$\phi = \tan^{-1}(X/Y)$$

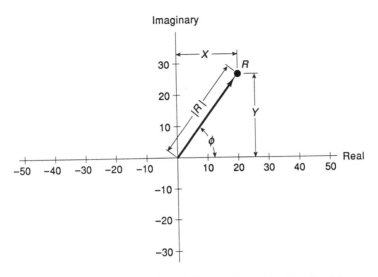

Figure 8.2. A representation of the complex point R = X + jY.

These relations can be seen in Figure 8.2. Note that

$$R = X + jY = |R| < \phi$$

For some of the functions, it is necessary to define the complex conjugate of R, or R^* (see Figure 8.3), as

$$R^* = X - jY = |R| < \phi$$

Some of the other complex-value notations that we will be using for the descriptions of the various functions include the following.

1. The linear magnitude of a function is $|H_{ab}|$
2. The logarithmic magnitude is $20 \log |H_{ab}|$ for a linear property or $10 \log |H_{ab}|$ for a squared (power) property
3. The phase shift is $\phi_b - \phi_a$
4. The real part of the function is $|H_{ab}| \cos(\phi_b - \phi_a)$
5. The imaginary part of the function is $|H_{ab}| \sin(\phi_b - \phi_a)$

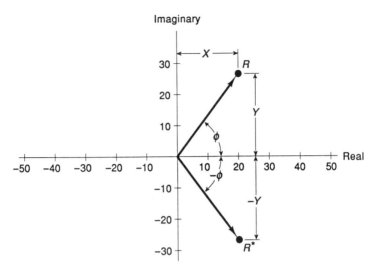

Figure 8.3. A representation of the complex points R = X + jY and R* = X − jY

This brief description of the manipulation of complex values will help in understanding some of the following function definitions. It should be remembered that these definitions represent the way in which the spectrum analyzer processes the sets of 400 amplitudes and phases for frequency domain properties or the 1,024 time domain points (assuming a 400-line spectrum analyzer). There are other ways to calculate some of these properties that make more sense to human one-number-at-a-time processors.

Input Time

The input time, or time domain display, of the FFT analyzer was described previously as what one would see on an oscilloscope. This is not exactly correct if one is thinking of a regular, analog oscilloscope. There are several reasons for this, any one of which can give the user the wrong impression of the phenomenon being observed:

1. Since the time display shows only digitized data, only the frequency content of the signal of interest in the selected analysis range being measured will appear. It is especially important to realize this in the case of a zoom analysis band, which may well eliminate the major low-frequency elements of the signal.

2. The signal may be clipped by the improper setting of the attenuators, or there may be a dynamic range problem as described in Chapter 3.

3. As described in Chapter 2, a 400-line spectrum analyzer with a sampling rate of 2.56 will have 1,024 time data points. Several of the more modern 400-line analyzers use a video display capable of showing only 400 points. On such analyzers, it is only possible to see part of the time display at any one time, making it necessary to scan through the display to see the complete time history.

4. Remember that the input time history represents only a single, unaveraged, memory period of the analyzer. If the time history of a nontransient nature is important, it is often desirable to time average as described in Chapter 2 in order to obtain some statistical accuracy.

The notation for the input time is $A(t)$ or $B(t)$ and is useful for observing transient phenomena. As mentioned in Chapter 5 concerning the acquisition of Transfer Functions, the viewing of the input time history after each hammer blow of an impact test is quite useful for determining if a good blow and ring down were obtained for each of the averages taken during the test. These observations will be useful in the event that the coherence for the test turns out to be less than desired.

One method of determining the amount of damping in a system is to measure the log decrement of the ring down. This is done by measuring the amplitude of two successive peaks in the time history ring down, as shown in Figure 8.4 and using the equations for log decrement and, if desired, critical damping.

$$\text{log decrement} = \delta = L_n \frac{X_n}{X_{n+1}}$$

$$\text{critical damping} = \zeta = \frac{\delta}{4\pi^2 + \delta^2}$$

In analyzers employing a large memory buffer zoom, it is possible to "catch" an event resultant from some impulse even if the exact timing of the event with respect to the impulse is unknown. For example, if one selected a large memory buffer zoom with a decimation ratio of 10 (10 × better frequency resolution) in the frequency range

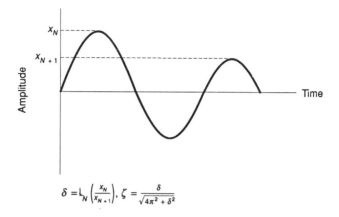

$$\delta = L_N\left(\frac{x_N}{x_{N+1}}\right), \zeta = \frac{\delta}{\sqrt{4\pi^2 + \delta^2}}$$

Figure 8.4. Log decrement.

0–500 Hz, the memory period would be 10 times the usual memory period for that range of 400 filters/500 Hz = 0.8 sec, or 8 sec. Thus, one could capture the event up to 8 sec after the impulse and sweep through the time display until the event is found.

Instantaneous Spectrum

The instantaneous spectrum is the complex spectrum of the input time data, or

$$S_a = F[A(t)]$$
$$= |S_a|\cos(\phi_a) + j|S_a|\sin(\phi_a)$$
$$= \text{real} + j\text{ imaginary}$$

also, log magnitude = $20 \log S_a$.

Note that the instantaneous spectrum is calculated as in Chapter 2 and is complex in nature, that is, has real and imaginary parts. In a single-channel analyzer, the real and complex parts of the spectra are used to calculate the magnitudes of the amplitudes of each of the 400 frequency cells and the phases are usually thrown away. This is because there is no reference point that can be used from which the phases may be measured.

The only exception to ignoring phase in a single-channel spectrum analyzer is in the case of synchronous time averaging. Since, in this case, the starting point of the memory wheel is well defined and corresponds to a particular physical event (such as the passing of a keyway), it makes sense to save the phases.

In a dual-channel analyzer, the phases of each of the 400 lines of each channel are saved. The meaningful parameter then becomes the phase difference between the channels in the form of the average cross-spectrum function mentioned later in this chapter.

Many people use the instantaneous spectrum far too often. It is important to remember that the instant FFT contains both the repetitive and random components of the signal. As such, there is no statistical accuracy to the amplitudes in each frequency bin. For this reason, averaging should almost always be used rather than the instant spectrum. If a nonstationary signal is being watched, consider taking a low number of exponential averages to gain at least some accuracy rather than using the instant FFT.

Power Spectrum

The power, or average, spectrum is the average of the squared magnitude of a number of successive instantaneous spectra. The average spectra has only a real part because the square of the imaginary part of the instant spectra is real (since $j \times j = \sqrt{-1} \times \sqrt{-1} = -1$, which is a real quantity).

The calculation of the power spectrum is

$$G_{aa} = \overline{(S_a \times S_a)} = \overline{(S_a^2)}$$

Note that the power spectrum is a squared quantity, with units of, say, g squared or (in/sec) squared. The log magnitude of the power spectrum, therefore, is

$$\log \text{ magnitude} = 10 \log G_{aa}$$

Since most analysts are more comfortable using a given set of linear units rather than units squared, they often view the power spectrum but read out in rms units, or view the rms averaged spectrum, given by

$$\text{rms spectrum} = \sqrt{G_{aa}}$$

Because of the way in which the rms spectrum is calculated, via the power spectrum, the rms spectrum is also a real display of rms amplitude versus frequency.

Power Spectral Density

Many FFT spectrum analyzers have a button labeled "Power Spectral Density," "PSD," or "$(\ )^{1/2}/\sqrt{Hz}$." The first three of these are the same function, the last is simply the square root of the others. To calculate the PSD of a spectra, the analyzer simply divides the amplitude of the measured power spectra at each frequency by the bandwidth of each frequency bin in Hertz. In the case of the $RMS/\sqrt{Hz}$. button, the square root of this value or the rms spectra divided by the square root of the bandwidth is calculated.

The purpose of PSD is to try to make some sense of wideband noise. Limits can then be set on the amount of vibration or sound level a machine exhibits when the cause is the general wideband noise found in nature. The following should explain the problem.

Suppose one has a pure tone of, say, 100 dB of sound power at 50 Hz. If this peak were measured using a 400-line FFT spectrum analyzer in the analysis range of 0–200 Hz, the amplitude of the signal would be 100 dB in the filter bin whose center frequency was 50 Hz. If the same signal was measured in the analysis range of 0–100 Hz, the result would still be a 100 dB amplitude in the 50-Hz frequency bin. This is because a pure tone at 50 Hz has all of its energy at 50 Hz and at no other frequency. (Note that we are assuming an ideal case—no frequency drift, a cell centered signal, etc.)

Suppose now that one had a white-noise signal, which was measured as having a power spectrum amplitude of 100 dB in the 50-Hz frequency bin (obviously, by the definition of white noise, all of the other bins also would show an amplitude of 100 dB) of an analyzer whose analysis range was set at 0–200 Hz. If this same signal were measured with the analysis range changed to the 0–100 Hz, the amplitude in the 50-Hz bin would be 97 dB. Why?

Since white noise has a uniform energy distribution across all frequencies, if we read the same signal with a filter bin half as wide, only half as much energy will reside in that bin. Remember that for power measurements, a reduction of energy by one half is equivalent to a 3-dB drop in amplitude (10 log 0.5 = –3). Unlike the case of a discrete spike as above, the amplitude of a white-noise measurement is a function of the filter width used.

One way around the problem described above is to define the amplitude on a per-Hertz basis. A white noise signal with an amplitude of 100 db/Hz in any of the 400 frequency bins on an analyzer set to the 0–200 Hz analysis range (with uniform weighting) would show the same 100 db/Hz amplitudes in PSD mode as in the 0–100 Hz range. On a non-PSD basis, however, the 0–200 Hz readings for a 100 db/Hz signal would be 100 db/0.5 Hz = 103 db (a doubling of power). The same 100 db/Hz signal in the 0–100 Hz range would be 100 db/0.25 Hz = 106 db for the non-PSD case. The amplitudes would depend on the bandwidth of the filters being used (as well as the shape of the weighting window used).

If PSD were used to measure the pure tone at 100 Hz described above, its amplitude for the 0–200-Hz analysis would be 100 dB divided by 0.5 or 2 × 100 dB/Hz = 103 dB/Hz. In the 0–100-Hz range, the amplitude would be 100 dB × 4 = 106 dB/Hz. The same spike would have different values depending on the analysis range.

It should be obvious that, if pure tones are of concern, normal amplitude readings should be used. If the concern is white noise, PSD should be used. If the signal to be measured is comprised of neither pures tone nor pure white noise, as machinery signals often are, a judgment must be made as to what the spectrum most look like.

There is another confusion concerning PSD, which often goes unnoticed by the technician because the spectrum analyzer, if it is properly designed, takes care of the problem automatically. The issue will be discussed here because it must be taken into consideration by anyone trying to implement PSD in a computer. The problem is the effect a particular weighting window (see Chapter 2) has on the value of PSD amplitude.

If white noise is input to a spectrum analyzer using an unweighted time window, each of the 1024 time history bins will contain the amplitude of the noise that existed when the digitizer "sliced" a particular piece of data. Thus, the FFT will be performed on that time window assuming constant white noise throughout the time increment. The amplitudes in the frequency domain will therefore be correct, and the division by the bandwidth of each of the frequency cells will yield the correct values of PSD.

Suppose, however, that the white noise is being fed into the analyzer via a Hanning time window. The energy of the noise residing at the beginning and ending of each memory period will be reduced because of the shape of the time window. When performing the FFT, the analyzer will think that there is less energy in the signal than there actually is. An amplitude error in the frequency domain will result. Note that this same kind of error results when a pure tone is operated on by a Hanning window (depending on the relation between the signal period and the memory period). Unlike a pure tone, white noise has the property that it has no period. Therefore, the error caused by weighting can be accurately predicted and a correction factor can be implemented in the PSD calculation to correct for the reduction of energy at the end points of the memory period. For a Hanning window, one obtains the correct value of PSD by dividing by 1.5 × bandwidth rather than by simply dividing by the bandwidth. For a Hamming weighting, one divides by 1.4 × bandwidth.

For some reason, unfathomable to the author, some military specifications call for PSD and some call for $[\ ]^{1/2} / \sqrt{Hz}$. There seems to be a high probability that the button on your spectrum analyzer will not correspond to the specification you must meet. This is because most analyzers simply divide the spectrum displayed by either Hz or $\sqrt{Hz}$ without giving any thought to what makes sense. Until this problem is corrected by spectrum analyzer manufacturers, one must read the manual and square or square root the results by hand or by computer.

Average Cross Spectrum

The average cross spectrum is the most basic of all dual-channel functions. Many dual-channel spectrum analyzers do not permit the user to view the average-cross-spectrum function because, in and of itself, it has no useful analysis purpose. It is used to calculate all of the other dual-channel functions because it is the most basic function retaining phase shift information between channels A and B.

The average cross spectrum is calculated as the complex product of spectrum S_b and the complex conjugate of S_a.

$$G_{ab} = \overline{(S_b \times S_a^*)}$$

$$= |\overline{S_a}|\,|\overline{S_b}|\,\cos(\overline{\phi_b - \phi_a})$$

$$+ j\,|\overline{S_a}|\,|\overline{S_b}|\,\sin(\overline{\phi_b - \phi_a})$$

$$= |G_{ab}|\,\cos(\overline{\phi_b - \phi_a}) + j\,|G_{ab}|\,\sin(\overline{\phi_b - \phi_a})$$

$$= \text{real} + j\,\text{imaginary}$$

The magnitude of the average cross spectrum indicates those frequencies where both $A(t)$ and $B(t)$ have content. The phase display indicates the frequency by frequency phase shift $\phi_b - \phi_a$.

Transmissibility

Transmissibility is the real ratio of the output power spectrum to the input power spectrum, or

$$H_{ab}' = G_{bb}/G_{aa}$$
$$\log \text{ magnitude} = 20 \log H_{ab}'$$

Transmissibility is fundamentally different from transfer function in that there is no concept of phase relationship in transmissibility. Therefore, one cannot gain an understanding of system resonances, zeros (anti-resonances), mode shapes, or damping using transmissibility. Coherence cannot be used to check the validity of a transmissibility test.

Furthermore, it should be noted that technicians who own only a single-channel analyzer cannot in any way simulate transfer-function tests by obtaining an output power spectrum, an input power spectrum, and asking the analyzer (or computer) to divide one by the other on a frequency-by-frequency basis. The very best that could be approximated is a transmissibility function, and then only if the data is very steady so that the time variation between the taking of the output spectra and the input spectra has little effect.

Transmissibility is used primarily to show things like sound attenuation versus frequency for sound-absorbing materials. It has little use in other applications.

Transfer Function

The transfer function has been dealt with in considerable detail in Chapter 5. It is probably the most useful dual-channel measurement, since it is used for determining the natural frequencies of a structure, mode shapes, nodes, antinodes, and structural damping, and is the basis of modal analysis and operating deflection analysis. Because of its importance, it is worth a few extra paragraphs here.

The transfer function defines the gain and phase lag of a system excited by $A(t)$ and responding with $B(t)$. It differs from transmissibility in that it is the complex ratio of the output spectra to the input spectra. For calculation purposes in a spectrum analyzer, it is defined as the ratio of the average cross spectra to the input power spectrum. This is because the cross spectra, as discussed above, is the most basic 2-channel property-preserving phase. If the reader will accept some rather sloppy mathematics, the equivalence of these definitions will be made apparent:

$$\text{transfer function} = H_{ab} = (\overline{G_{ab}})/(\overline{G_{aa}})$$

$$= (\overline{S_b \times S_a^*})/(\overline{S_a \times S_a^*})$$

$$= (\overline{S_b})/(\overline{S_a})$$

$$= \text{output/input}$$

The transfer function may be displayed in as many ways as complex numbers may be displayed. The most common way is a plot of the magnitude of the transfer function versus the frequency,

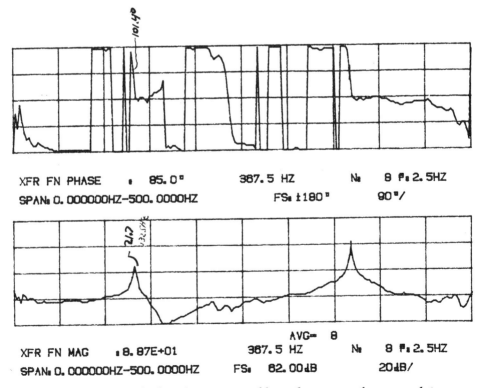

Figure 8.5. *A transfer function represented by a phase versus frequency plot and a magnitude versus frequency plot.*

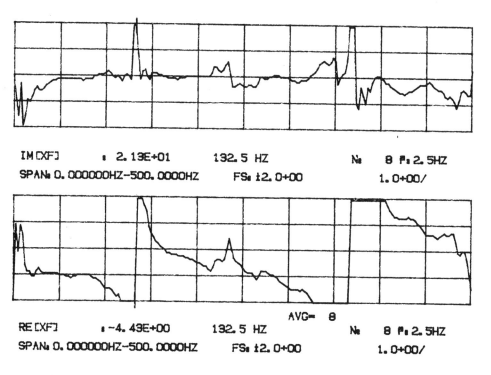

Figure 8.6. The same transfer function as in Figure 8.5 represented by an imaginary versus frequency plot and a real versus frequency plot.

accompanied by a plot of the phase shift versus the frequency, as shown in Figure 8.5. If both the frequency and amplitude axes of this display are logarithmic, the plot would be the classic Bode plot used by circuit designers and automatic controls specialists to describe the characteristics of their designs. For most of the purposes of the spectrum analysis diagnostician, however, a logarithmic frequency axis is of little use.

Another way to display the transfer function is with a plot of the real part of the complex amplitude versus the frequency and the imaginary part of the complex amplitude versus the frequency. These plots are shown in Figure 8.6. Note that if a force hammer and accelerometer are used, the imaginary value peaks at resonance and the real part goes through zero. If a velocity pickup is used instead of an accelerometer, the roles of the two plots for identifying a resonance is reversed because of the additional 90° phase lag between the velocity pickup and the accelerometer.

A third way to represent the transfer function is via a polar plot. Here, each value of magnitude is plotted from the origin at its corresponding value of phase to the horizontal axis. Frequency varies along the locus of these points (see Figure 8.7).

Although the polar plot of Figure 8.7 is of little help in picking out natural frequencies by eye, it lends itself well to the curve-fitting algorithms used for finding natural frequencies and mode shapes in modal analysis. Also note that this form of plot is called a Nyquist plot on some spectrum analyzers. Keep in mind that the plot we usually obtain in our search for natural frequencies has nothing to do with the Nyquist stability criterion usually thought of in relation to Nyquist plots.

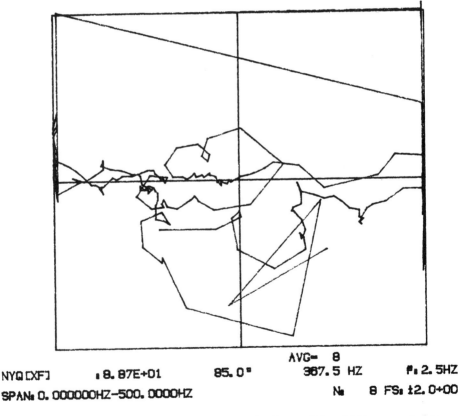

Figure 8.7. The same transfer function as in Figures 8.5 and 8.6 represented as a polar plot.

Force and Response Windows

Two special time windows are occasionally available on dual channel analyzers to simplify impact tests for transfer function (see Figure 8.8).

Input Force Window

The input force window multiplies the 1,024 pieces of input time data in channel A by zeros at all locations except for a few bins surrounding the actual impact. This fools the analyzer into ignoring superfluous input data such as double hammer blows in the same memory period or excessive background noise. As discussed in the section on inverse transfer function later in this chapter, this will have the effect of improving coherence. It is important to keep in mind that, even though the analyzer fails to pickup the extraneous excitation, the structure being tested may be affected by it.

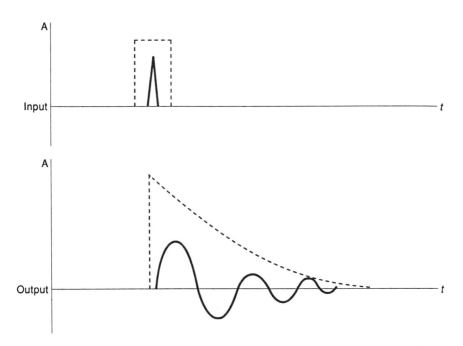

Figure 8.8. Two special time windows often used to simplify impact testing for transfer functions.

Response Window

The response window is an exponentially decaying function set to begin its decay after the first perturbation of the response of the structure and go to zero at or near the end of the time window. The purpose of this window is to fool the analyzer into thinking that the time window chosen for the test (which results from the analysis range chosen) was correct for the amount of damping in the structure. Thus a good ring down is obtained in spite of the relationship between the analysis range and the system damping. Keep in mind, however, that any damping determined from the "Q" of the peaks in the magnitude of the transfer function (see Chapter 5) will be wrong unless the artificial damping applied via the response window is taken into account. Realize, further, that even though this window fools the analyzer into thinking that the structure "rings down" in one time window, the structure is not deceived. Continuous ringing from a previous impact hammer blow will still cause poor coherence in a transfer function test.

Coherence

Coherence is probably the most valuable dual-channel FFT function besides transfer function. This is true because of its ability to measure the "goodness" of a transfer-function test, as well as because of its ability to help answer the question "who's doing what." If the reader understands the material on coherence, which was discussed rather fully in Chapter 5, only a few more points need to be covered here.

Coherence, which measures the linear relationship between $A(t)$ and $B(t)$, is calculated as

$$\text{coherence} = \frac{\overline{G_{ab}} \times \overline{G_{ab}}^*}{\overline{G_{aa}} \times \overline{G_{bb}}}$$

Remember that, in the case where the $A(t)$ signal to the analyzer represents an excitation (cause), and the $B(t)$ signal an effect, a coherence of 1.00 indicates that all of $B(t)$ was due to $A(t)$. If, on the other hand, $A(t)$ is also an effect, then a coherence of 1.00 simply indicates that both $A(t)$ and $B(t)$ are seeing nothing except the same external excitation source.

To further explain this point, a short case history is useful. The author was asked to determine the reason for occasional excessive vibration at the motor of a rather large centrifugal vacuum pump that extracted steam from a process. Three things were found:

1. The occasional vibration was all at 50 Hz, which was the running speed of the vacuum pump (the motor ran at 60 Hz through a gearbox)
2. The base had a natural frequency of approximately 50 Hz
3. During the periods of high vibration, a pressure transducer in the volute of the pump had a high amplitude and was approximately 0.90 coherent with the accelerometer mounted on the idle end of the motor

Since both the accelerometer signal and the pressure transducer signal are "effects" rather than "causes," one could not prove that the pressure was exciting the idle end of the motor via the resonance or vice versa. Since the pump was quite large and constructed of heavy gauge stainless, one had to reason that it was too big to be an efficient diaphragm pump, and that the hydraulics inside the pump had to be causing both the pressure pulses and the motor vibration at 50 Hz.

It is important to realize that coherence is calculated by manipulating arrays of averaged properties. If one were to display the coherence function while taking a single average of an impact test, the coherence would be 1.00 at all frequencies. This does not mean that the test is perfect, only that coherence is not defined in the absence of averaging.

The coherence function defined above, and as it is available in any spectrum analyzer, deals with linear relationships. It may well be that a particular cause is indeed causing a particular effect, but the relation is represented by some polynomial, rather than linearly. Thus, the coherence may be 0.00 even though the cause is, indeed, causing the effect. In such a case, it is possible to define a bicoherence for square relationships, a tricoherence for cubic relationships, and so on. These functions are so specialized, however, that they will not be found on any commercial FFT spectrum analyzer.

The author is often asked what level of coherence a technician should accept as being an indicator of a reliable impact test. The answer is always the same: "Who knows?"

Certainly, a coherence of 0.80 or better is usually a good bet. A coherence of 0.20 or less must be viewed with great skepticism. The author will often accept coherences of 0.50–0.80, if several sets of impact tests have been run with no improvement in coherence, and there is an understanding of why a better test cannot be run under prevailing conditions.

Since one of the most important uses of the coherence function is to indicate the reliability of a measured transfer function, many dual-channel analyzers offer the ability to do coherence blanking. The user decides what value of coherence is acceptable as an indicator of a sufficiently valid test and inputs this into the spectrum analyzer via its keyboard. The analyzer will then display and/or plot out only those parts of the transfer function plots that corresponds to frequencies with coherence values greater than the user-specified limit. The areas where the coherence is too low are left blank.

Coherent Output Power

The coherent output power was mentioned briefly in Chapter 5 and will be defined just as briefly here. It is simply either the power in signal $B(t)$ that was caused by the power of signal $A(t)$ or the power of signal $B(t)$ that is being excited by the same phenomena that is exciting $A(t)$. The correct definition depends on whether one is dealing with a linear cause/effect relationship or not, as discussed above for transfer functions.

For completeness, the equation for coherent output power (COP) is, simply,

$$COP = \text{coherence} \times G_{bb}$$

and (since COP is a power measurement)

$$\log \text{magnitude} = 10 \log COP$$

See Figure 8.9.

Impulse Response

One of the mathematical reasons that working in the frequency domain seems simpler than working in the time domain is that integration in the time domain is equivalent to multiplication in the frequency domain. If one has determined the frequency response

Advanced Analyzer Functions **187**

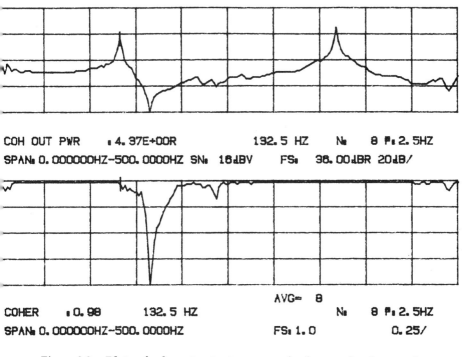

Figure 8.9. *Plots of coherent output power and coherence for the transfer function test of Figures 8.5–8.7.*

of a linear system and knows the amplitude/frequency characteristics of the excitation of the system, it is a simple matter to predict what the output of the system will be. The equation is simply

$$\text{output} = \text{transfer function} \times \text{input}$$

In the days before FFT spectrum analyzers and computers, it was not a trivial matter to determine the frequency characteristics of the input signal or the transfer function of any but the most simple-minded systems. The problem of predicting an output was often solved in the time domain.

To do this, one calculated the response of a system to a unit impulse (a mathematical fiction). The answer arrived at by this arduous task was called the impulse response.

Once the impulse response was known for a given system, the known excitation was approximated by a series of scaled impulses offset by time delays from each other. The output of the system due to each impulse could be found simply by multiplying the impulse response by the scaling factor.

The results of each of the set of calculations mentioned in the previous paragraph was then added together (adjusting for time delays) to obtain the full response to the full excitation. This was done using the convolution, Faltung, or DuHammel's integral (three names for the same thing).

The calculation of responses was a time-consuming task performed by hand. Its implementation in a computer algorithm is hampered by the difficulty of working with continuous integrals in computers. The task of multiplying an array of 400 discrete amplitudes versus frequencies of a forcing spectra by an array of 400 magnitudes and phases versus frequencies is easily handled by an FFT spectrum analyzer (which is basically an array processor with special filtering).

There are two advantages of solving the input/output problem using the impulse response time domain. They are both based on the fact that humans tend to think in the time domain, not the frequency domain.

1. The impulse response, which is a plot of amplitude over time, gives an intuitive picture of how the system reacts over time to an impulse. This makes it possible for an analyst to check his results with reality every so often.

2. Since the impulse response appears as a "ring down" for second-order systems (such as ones comprised of springs, dampers, and masses) and since it is arrived at in the dual-channel analyzer via averaging for statistical accuracy, it is a good way to determine damping via the log decrement method mentioned above with respect to Input Time (which is not averaged except for synchronous time averaging).

The impulse response is a dual-channel property, since it involves both an input and an output. Since it represents the response a system would have if the input had been a pure impulse (i.e., normalized), its implementation in a spectrum analyzer should clearly be the inverse FFT of the transfer function (which is the response normalized by the actual input). The equation is

$$\text{impulse response} = F^{-1}(H_{ab})$$

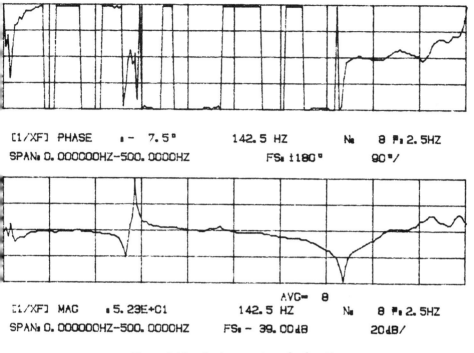

Figure 8.10. An inverse transfer function.

A typical display of an impulse response can be found in Figure 8.10 along with the transfer function magnitude.

Inverse Transfer Function

The inverse transfer function is a rather new function. It was defined to deal with the problem of obtaining an incorrect value of transfer function in the frequency range near a natural frequency caused by instrumentation noise.

The definition of transfer function, as shown previously, is

$$H_{theory} = G_{ab}/G_{aa}$$

or, cross spectrum divided by input spectrum. Note, however, that random noise averages out in the cross spectrum because there will not be a commonality between the noise in channel A and the noise in channel B. The actual signal of channel A, the input channel, will be comprised of both the signal of interest and noise that is not completely eliminated through summation averaging. Therefore, the actual equation for transfer function becomes

$$H_{actual} = G_{ab}/(G_{aa} + G_{nn})$$
$$= H_{theory}/(1 + G_{nn}/G_{aa})$$

Thus, the transfer-function values will be too high. Further, the coherence will be low because of the input noise

$$\text{coherence}_{actual} = \frac{G_{ab} \times G_{ab}^*}{(G_{aa} + G_{nn})(G_{bb} + G_{mm})}$$

where G_{nn} and G_{mm} are the noise inputs to channel A and channel B, respectively.

To deal with this problem, inverse transfer function has been defined as

$$H_2 = G_{bb}/G_{ba} = \text{output power/cross spectrum}$$

Since the actual output power spectrum is comprised of the signal of interest plus noise, the inverse transfer function becomes

$$H_2 = (G_{bb} + G_{mm})/G_{ba}$$
$$= H_{theory} \times (1 + G_{mm}/G_{bb})$$

Again, the measured inverse transfer function will have values that are too high. The advantage of using the inverse transfer function is that the only contamination is due to the noise in the output signal, the input noise does not show up in the calculation. Note that, in the area near a resonance, the actual output power spectrum is likely to be of a much higher amplitude than the instrumentation noise supplied to channel B ($G_{mm} << G_{bb}$). Therefore, the value of the actual inverse transfer function will be much closer to the theoretical value in the region of a resonance. A plot of the inverse transfer function of the test shown in Figures 8.5–8.7 is shown in Figure 8.10. A plot of the coherence is in Figure 8.11.

Note that the peaks (natural frequencies) of Figure 8.5 show up as valleys in Figure 8.10. The inverse of a peak is a valley.

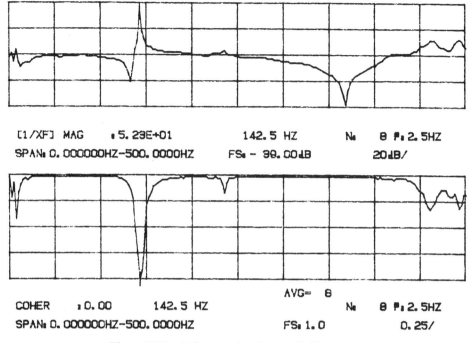

Figure 8.11. Coherence for the test in Figure 8.10.

Summary

Much of the information contained in this chapter will soon be forgotten by the reader. The important thing to remember is that you have seen these definitions in this chapter so that you can look up a specific item when a measurement problem rears its ugly head and you need help.

CHAPTER NINE

More Advanced Analyzer Functions

Introduction

In Chapter 8, we covered those advanced analyzer functions that might reasonably be used in the diagnostics of particular sets of problems. The functions discussed in this chapter are available on many modern FFT spectrum analyzers, but are rarely used. They were put into a separate chapter in order that a diagnostician could refer to them when the normal avenues of attack do not work on a particular problem.

Autocorrelation

Both the autocorrelation and the cross-correlation functions are time domain properties. Since they turn out to be inverse FFTs of frequency domain properties, their calculation can be easily implemented in an FFT spectrum analyzer. The autocorrelation function is a single-channel property; the cross-correlation function, described below, requires a dual-channel analyzer for implementation.

The autocorrelation function, which compares the similarity of a time domain function $A(t)$ to replications of itself delayed by a time constant τ, is given by the equation

$$R_{aa} = F^1 |G_{aa}| = \frac{1}{P} \int A(t) \times A(t - \tau)dt$$

where P is the period of the signal. Note the difficulty one would have in calculating the property in the time domain (integration) compared to the ease with which it can be processed in the frequency domain (simply an inverse FFT of the averaged power spectra). The fact that the averaged power spectra is used should be a hint that the autocorrelation function, like coherence, is an averaged rather than an instantaneous function.

One of the main uses of the autocorrelation function at the present time is to check a signal for periodicity. Suppose one were setting up instrumentation to make a measurement, which was thought to contain a square wave, but a look at the averaged spectrum of the signal yielded white noise (and, to further complicate the matter, there was no way to get a trigger signal for synchronous time averaging). Either it was incorrect to assume that the signal should contain a square wave, or the experiment was so noisy that the square wave was buried in noise. The autocorrelation function can solve this problem.

Since noise is random, it never repeats itself. Therefore, the autocorrelation of white noise shows information at time $t = 0$, but no correlation elsewhere (see Figure 9. 1). If a square wave was buried in

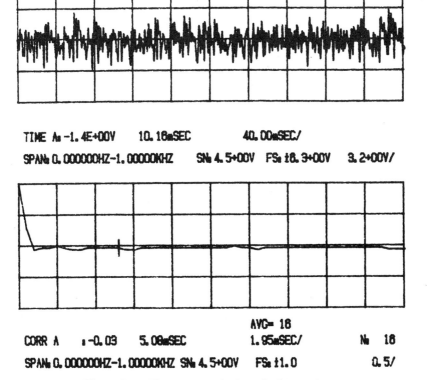

Figure 9.1. *The autocorrelation of white noise.*

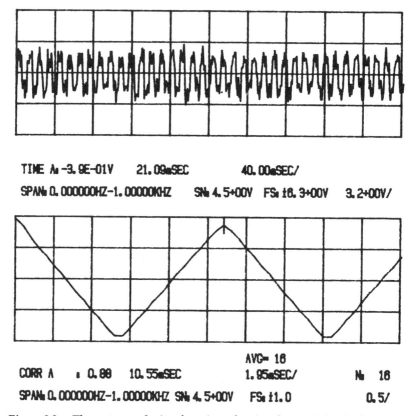

Figure 9.2. The autocorrelation function of a signal containing both a square wave and white noise.

the noise, it would have good correlation to itself at intervals equal to the period of the wave. Thus, if a square wave existed in our experiment, the autocorrelation would appear as in Figure 9.2, a square wave plus noise. We would soon know if our instrumentation was faulty (too much noise) or if our assumptions about the nature of the signal were incorrect. Figures 9.3 and 9.4 show the autocorrelation functions for a square wave and a sine wave, respectively.

An interesting application of the autocorrelation function with respect to the condition of gears and rolling element bearings has been used in several different applications with some success. Consider the case of a set of perfect gears meshing. Each event, comprised of a perfect tooth on one gear meshing with a perfect tooth on the other gear, should look the same as the next meshing event caused by the next pair of perfect gear teeth. We should then expect

More Advanced Analyzer Functions   195

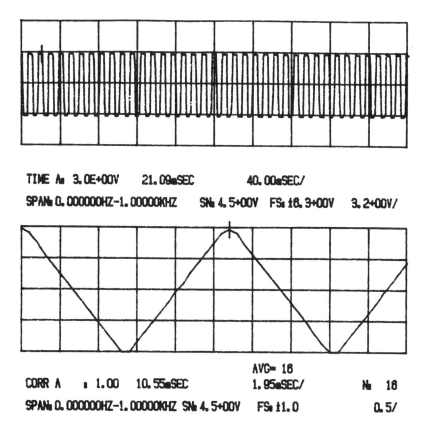

Figure 9.3. The autocorrelation function of a 94.8-Hz square wave.

a well-defined autocorrelation function with a period equal to the inverse of the mesh frequency. In fact, if one knew the shape of the profiles of the teeth, one could predict the shape of the autocorrelation function. Any deterioration in the quality of the meshing of the fictional perfect gears would show up as a change in the shape of the autocorrelation function taken over a revolution of the gears.

Since real gears come with off-the-shelf errors, the use of the autocorrelation function to check on either the "goodness" of new gears (degree of imperfection) or for deterioration of old gears, one would have to develop baseline criteria. These could be developed mathematically for new gears, or from past measurements as a predictive monitoring tool.

Obviously, any repetitive event could lend itself to analysis via the use of the autocorrelation function. The main obstacle to such an analysis is the fact that most machinery analysts have already had significant success working in the frequency domain. It may be well to keep in mind that past satisfaction is no guarantee of future success.

196 *Chapter Nine*

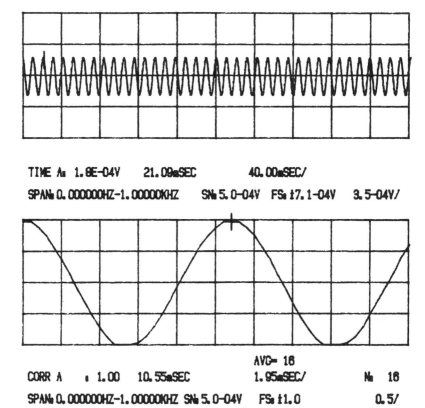

Figure 9.4. The autocorrelation function of a 94.8-Hz sine wave.

Cross-Correlation

As mentioned above, cross-correlation is a 2-channel time domain property. Like autocorrelation, it is calculated in a spectrum analyzer as an inverse FFT. In this case, the frequency domain property used is the cross spectrum, the primary dual-channel property. Cross-correlation is useful for obtaining reaction times in a process and in finding process flow rates.

The cross-correlation function compares the similarity of $A(t)$ and $B(t)$. The value of time, τ, which maximizes the value of the cross-correlation function yields the time lag between the two signals. The equation for cross-correlation is

$$R_{ab} = F^{-1}[G_{ab}] = \frac{1}{P} \int_O^P A(t) \times B(t-\tau)dt$$

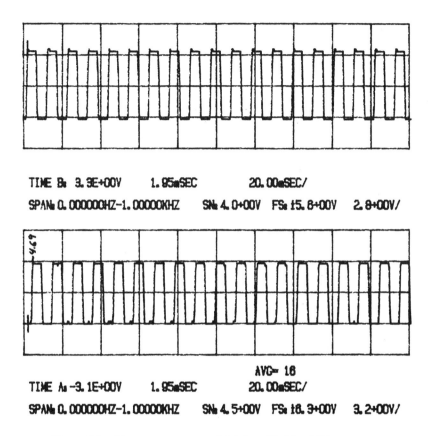

Figure 9.5. Two square waves A(+) lags B(+) by 2.74 m sec.

Figure 9.5 shows the time traces of two square waves. The lower signal lags the upper signal by 2.74 msec. Figure 9.6 shows the cross-correlation function between the two signals. At the cross-correlation value of 0.99, the time can be read as 2.73 msec.

Figure 9.7 shows a way to use cross-correlation to determine the velocity of a slurry in a pipe. Two pressure transducers are mounted a distance L apart in a pipe. The cross-correlation between channels A and B of a dual-channel spectrum analyzer will yield the time delay between the two pressure transducers. The velocity of the flow is given by L/τ. A similar method can be incorporated between sonar buoys in the ocean to triangulate the location of an enemy submarine.

Figure 9.8 shows a test one can employ to determine the time delay of a motor relay. Channel A of the spectrum analyzer monitors the power to the relay. Channel B sees the power to the motor (voltage reduced to keep from frying the analyzer). When the "off" button

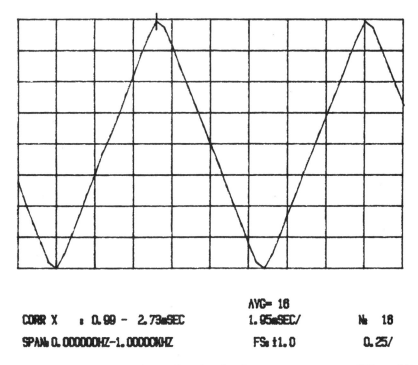

Figure 9.6. The cross-correlation function for the two square waves of Figure 9.5.

on the motor controller is hit, the relay loses power immediately. The motor will not lose power until the mechanical action of the relay takes place. The author has used this phenomena to sufficiently vent a compressor to allow for an easy motor shutdown. The same method can be used to determine the response time of devices such as solenoid valves and other coil driven equipment.

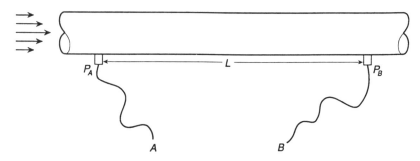

Figure 9.7. Slurry flow in a pipe.

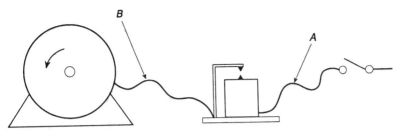

Figure 9.8. *Relay delay times.*

Probability Density/Cumulative Distribution Functions

Some spectrum analyzers are capable of doing a statistical analysis of the time data in the input memory buffer. The probability density function is the statistic defining the most probable amplitude of $A(t)$. Of more practical interest is the cumulative distribution function. This latter function defines the probability that a time signal, $A(t)$, has an instantaneous value of, say, X volts or less.

The probability functions can be useful in determining the characteristics of a given signal and, therefore, can indicate when an instrumentation problem exists. The following functions should be studied in order to get used to using the probability functions:

1. White noise. Notice the classic Gausian bell curve of the probability function in Figure 9.9.

2. Sine curve. The cumulative probability of having a voltage equal to or lower than the peak of the sine wave is 1.00. The cumulative probability of seeing a value of less than the negative peak of the sine wave is 0.00 (see Figure 9. 10).

3. Square wave. Essentially, the square wave only has energy at its maximum voltage and its minimum voltage; it has no energy in between (see Figure 9.11).

4. Square wave plus noise. This probability distribution has the two end peaks of the square wave plus a narrow bell curve in the center, as shown in Figure 9.12. If the noise level was higher with regard to the square wave, the bell curve would have been larger.

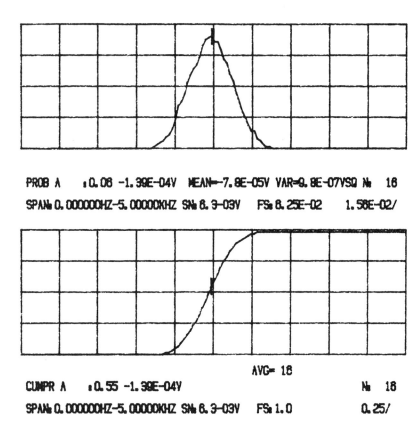

Figure 9.9. *The probability functions for white noise.*

In recent years, attempts have been made to describe the characteristics of a signal in terms of the quantities used to describe the probability distribution, or histogram of the data, particularly for the gears and bearings of helicopter drives. It is well known that the mean and standard deviation of measured data makes some basic statements about the probability distribution of the data. To completely describe the histogram of the data takes 99 different parameters. The mean is defined as

$$\text{mean} = \mu = \frac{1}{n} \sum_{i=1}^{n} X_i$$

where X_i represents one of n pieces of data. The variance, or square, of the more common term, standard deviation, is given by

$$\text{variance} = s^2 = \frac{1}{n} \sum_{i=1}^{n} (\mu - X_i)^2$$

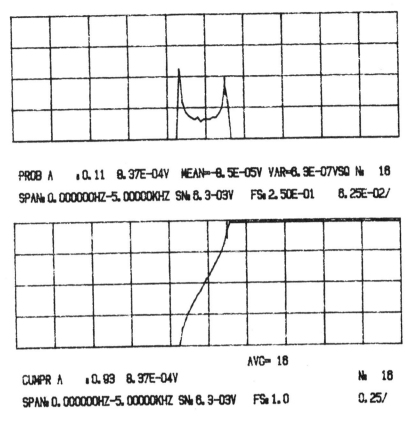

Figure 9.10. *The probability functions for a sine wave.*

The n_{th} moment describing some characteristic of the histogram is given by

$$n^{th} \text{ moment} = \frac{1}{n}\sum_{i=1}^{n}[(\mu - X_i)^n]$$

Note that, by this definition, the variance is the second moment. Some recent papers on helicopter vibration indicate that the fourth moment statistic of the vibration time waveform histogram is indicative of gear and bearing condition. A few papers find the sixth moment statistic to be of use.

Acoustic Intensity

Acoustic intensity measures a kind of vector sound pressure. This is useful for noise-source identification, where the system under

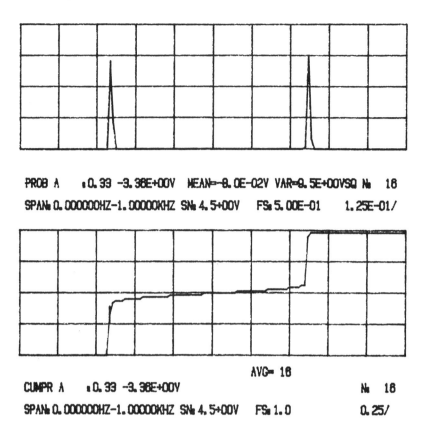

Figure 9.11. *The probability functions for a square wave.*

consideration is so stiff that the coherence is 1.00 between all points in the system. If a vector sound pressure were found over the surface of the system, the largest vector would point to the area radiating the most noise. This would identify the culprit.

In its most common embodiment, acoustic intensity is found using two carefully phase-matched microphones separated by a fixed distance. The output of each mike goes to one of the channels of a 2-channel FFT analyzer. This is shown in Figure 9.13. The acoustic intensity is then calculated as

$$I = (\text{imaginary part of } G_{ab})/(D \times \rho \times \omega)$$

where ρ is the density of air and (ω) is the frequency in radians per second.

If the value of acoustic intensity were integrated over the surface of a noise source, the result would be sound power. Sometimes, an acoustic intensity map, similar to a deflection map commonly seen

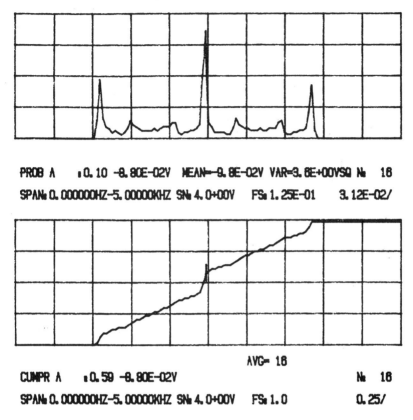

Figure 9.12. The probability functions of a square wave plus noise.

in modal analysis, is made. This map shows how the device being tested acts as a sound speaker generating noise.

A second way to determine acoustic intensity is to use vibrational velocity readings at the surface of the test object (to obtain the vector quantity) and the sound pressure level near the surface. This method removes the need to compensate for non-phase-matched microphones.

Cepstrum

There are few functions available in spectrum analyzers that have caused as much controversy among vibration analysts as cepstrum. Most analysts refuse to use this function even though it can present a less complicated plot (cepstra) than is available in the frequency domain (spectra). This may be because most analysts, the author included, are used to looking at spectra and are successful at solving problems in the frequency domain.

204 Chapter Nine

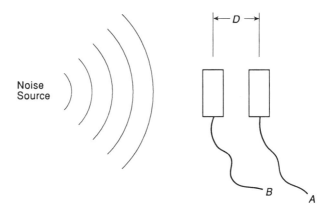

Figure 9.13. A schematic of the microphone setup used to measure acoustic intensity.

The cepstrum of a signal is defined as the inverse FFT of the log of the complex spectrum, or

$$C = \tilde{f}^{-1} (\log G_{aa})$$

The resultant plot, or cepstra, is on axes of amplitude versus Quefrency in milliseconds. Perhaps this silly name for a time axis is another reason for the reluctance of vibration analysts to use cepstrum.

An older definition of cepstrum was

$$C = |\tilde{f} (\log G_{aa})|^2$$

The most painless way to think of cepstrum is to realize that it is nothing but an autocorrelation with a logarithm in it. This means that it has the same potential uses described in the section on autocorrelation. The primary claimed advantage of cepstrum over autocorrelation is the fact that

$$\log (A/B) = \log A - \log B$$

This means that it is possible to edit the cepstra to simplify it. Unwanted terms can be edited out leaving, hopefully, the one kernel of information required to solve the problem at hand. The author has read many papers on the analysis of gear mesh problems that were solved using cepstra. Unfortunately, none of these papers showed a solution that could not have been found by zooming in on the sidebands and mesh frequencies of the gearbox.

It should be mentioned that cepstra is used in speech analysis. It can be done in using any spectrum analyzer capable of performing an inverse FFT by transporting the power spectra buffer in the analyzer memory to a computer, taking the logarithm on a point-by-point basis, and transporting it back to the analyzer for the inverse FFT operation.

Summary

Admittedly, the descriptions of the functions in this chapter were rather sketchy. This is because they are seldom needed. In some cases, such as autocorrelation, and the histograms, more research must be done. In cases such as cepstrum, the use of the function is small due to simple reluctance to use it. The reader is invited to realize that these functions exist and never to think of them again, unless the usual ways of looking at a problem fail and a new technique is needed.

APPENDICES

The following appendices cover several topics that, although they might be crucially important to diagnose a problem or to the conduct of a test at some time, would have badly muddled the basic concepts presented in previous chapters. Now that the reader understands the basic concepts, we can present more complex notions. The reader is invited to pick and choose among the appendices as needed.

Nobody knows everything. Over the years, the author has met people possessing tremendous experience in a particular area of interest. Some of these people have been prevailed upon to provide appendices to this second edition in the hope of providing more background information for the vibration analyst—a person who is expected to know everything about everything.

APPENDIX A

Reading Spectral Plots

Throughout this book, figures have been used that show actual data plotted on an FFT spectrum analyzer. The presentation of these plots could have been handled in one of two ways:

> The data could have been displayed with only enough notation to reinforce the point being made in the text. If this had been done, the reader looking at the data would have been left with the uncomfortable feeling that there was valuable information resident in each plot which could not be retrieved because of the insufficient notation. The resultant anguish might have caused some readers to totally miss the point of the figure.

> If, on the other hand, full notation was included with each plot, the more casual or inexperienced reader might lose the point of the plot amidst vast amounts of irrelevant notation and data. This reader might soon give up trying to understand the spectral plots.

Upon serious consideration, it was decided to add this appendix to the book to explain how one reads the spectral plots. In this way, the data seeker can properly look at the data while the searcher of concepts can know which pieces of information contained in the plots are not relevant.

It should be noted at this point that each manufacturer of spectrum analyzers tends to use different notation on their plots. The good news is that, once the technician has mastered reading the plots of any spectrum analyzer or monitoring device, it is an easy matter to figure out the notation of the plots generated by any other instrument.

208 Appendix A

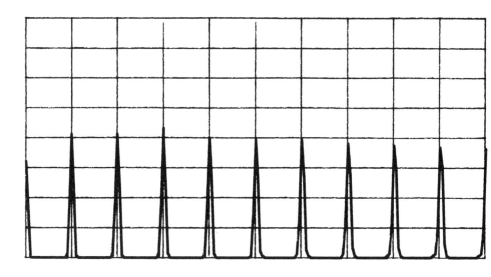

Figure A.1. Reading spectral plots.

Almost all spectral plots are shown in a Cartesian X-Y format in which the horizontal axis is in units of frequency such as Hertz, cycles per minute, or orders. The vertical axis is in amplitude units such as g's RMS, inches per second peak, or mils peak-to-peak. The text appearing above or below the plot is meant to inform the observer as to the kind of plot displayed, the value of the frequency and displacement at the cursor (usually a vertical tic mark, line, small rectangle, etc.), the scale of the curve, and whatever information the manufacturer thinks you need to know about the setting of the controls of the instrument at the time the data was taken. Figure A.1 shows a plot of the test signal of the analyzer used to generate the plots of this book. The meaning of each piece of text in the plot is as follows:

PWR SPECT A   The data shown is a power spectra taken on channel A of the dual channel analyzer.

X.XXE-XXR   The value of the amplitude at the cursor is $X.XX \times 10^{-xx}$ reference units. The reference units are added by the user in terms of the transducer used. See Chapter 3 on transducers.

XXX.XX HZ.	The frequency at the cursor in Hertz.
N: 8	Eight averages requested to arrive at this plot.
AVE = 8	Eight averages were actually taken to arrive at this plot.
β:XXHZ	The bandwidth of each of the analyzer's filters is XX Hz. (Note that Span/400 filters = XXHz.)
SPAN:0.0000-10.0000kHZ	The frequency span of the plot is 0 to 10 kHz. That is, the extreme left vertical line is zero Hz. and the extreme right is 10 kHz. Therefore, each vertical division is 1 kHz. wide.
SN:$X.X$–XXV	The sensitivity of the analyzer during the testing was set at $X.X \times 10^{-xx}$ volts.
FS:$X.X$-XXR	The upper horizontal line (full scale) has an amplitude of $X.X \times 10^{-xx}$ reference units.
$X.X$-XXR/	Each vertical division is $X.X \times 10^{-xx}$ in value.

The same meanings can be ascribed to a second plot displayed in conjunction with the plot shown. Some of the other notation used in this book is given in the table below.

XFR FN MAG	Transfer function magnitude. In this case, the amplitude at the cursor is a dimensionless ratio.
XFR FN PHASE	Transfer Function phase.
IM[XF]	A plot of the imaginary part of the transfer function. See Chapters 5 through 9 for a description of this and other dual channel properties.
RE[XF]	A plot of the real part of the transfer function.
NYQ[XF]	A Nyquist plot of the transfer function.

COHER   The coherence of the attendant transfer function plot. The amplitude of the coherence plot always varies between 0.00 and 1.00 coherence.

COH OUT PWR Coherent output power.

[1/XF]   This refers to the inverse transfer function.

Time domain plots differ from frequency domain plots in that the cursor values are in terms of an amplitude at some value of time such as milliseconds and the value of the horizontal divisions are given in terms of msec/division. Since the cross-correlation and autocorrelation functions are time domain plots, the format is similar to the amplitude *vs.* time plots.

Histogram and cumulative probability plots display a dimensionless value of probability at the cursor *vs.* a value of voltage. The vertical amplitude of the plots vary between a probability of 0.00 and 1.00.

With the above explanation of the plot notation used in this book, and a few minutes spent looking at some of the plots, it should become evident that similar plots generated by any instrument can be easily decoded. The problem is not in reading plots, it is in understanding what they mean.

APPENDIX B
Pulse Theory

This author has spent many years lecturing to different groups of people who were interested in spectrum analysis for problems ranging from detailed physical tests of cavitation phenomena to philosophical questions about why the walls of Jericho fell. Often, a very simple explanation involving pulse theory has sufficed to explain some of the observed phenomena. This section is designed to give the reader food for thought rather than quantitative answers.

The Basic Characteristics Of The Pulse

The most commonly discussed pulse is probably the impulse. The unit impulse is a mathematical fiction and is shown in the time domain representation of Figure B.1a as having an amplitude of 1 unit and a width of 0 sec. The Fourier transform of a single impulse is pure white noise in the frequency domain. Since white noise has equal amplitude at all frequencies, the energy of the impulse is spread rather thin, yielding a low amplitude of white noise at all frequencies (see Figure B.1b).

Since we know that nothing in nature happens over a zero time duration, it makes sense to talk about a pulse of finite width. Figure B.2a shows a pulse of time width T sec. Note that the frequency domain of the pulse looks very different from the pure white noise spectrum caused by an impulse. Examination of Figure B.2b shows a series of lobes in the frequency domain. The first lobe contains most of the total energy of the pulse and rolls off toward the zero amplitude line at a frequency of $1/T$. Each successive lobe has considerably less energy than the previous one and approaches zero amplitude at harmonics of $1/T$.

212 Appendix B

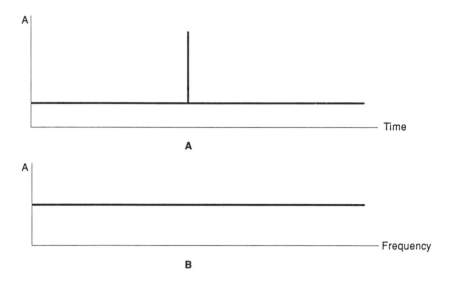

Figure B.1. The impulse.

Notice what happens when the FFT of a still wider pulse is taken, as in Figure B.3. Almost all of the energy of the pulse is located in the first lobe, going from 0 to $1/T$ in frequency (note that $1/T$ is a lower frequency than for the narrower pulse discussed above because T is larger). Succeeding lobes have relatively little energy.

Ramifications of Pulse Duration

The key to all that follows lies in comparing the plots for the three pulse widths of Figures B.1, B.2, and B.3. Note that, as the width of the pulse increases, the distribution of the energy caused by the pulse shifts to lower frequencies. The effect that this energy shift has in some areas of everyday life will now be discussed.

Impulsive Testing for Natural-Frequency Determination

As discussed in Chapter 5, one very convenient way to determine the natural frequencies of a structure is to do a dual-channel impulsive (or, more properly, pulsive) test. An accelerometer is mounted on a judiciously selected location of the structure in question and connected to channel B of a dual-channel spectrum analyzer.

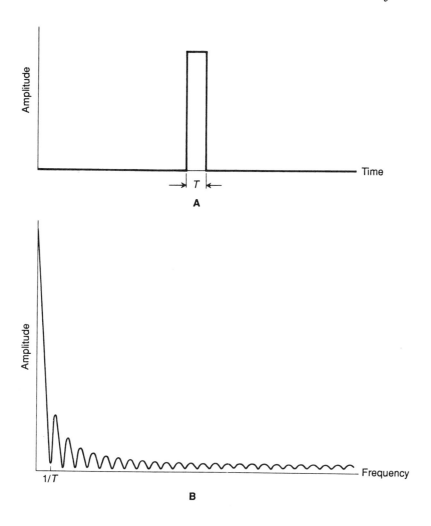

Figure B.2. A wide pulse.

A special hammer, with a built-in force transducer, is connected to channel A of the analyzer. The structure is hammered at certain locations a number of times. A set of averages is taken for statistical accuracy.

The transfer function is calculated by the analyzer as the cross spectra of A and B divided by the power spectra of A and coherence is checked to determine the accuracy of the test. If the coherence is low, the natural frequencies of the structure cannot be determined with adequate certainty. One possible cause of this poor coherence has to do with pulse theory.

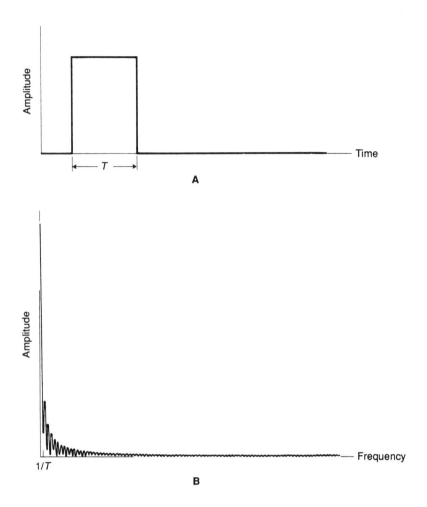

Figure. B.3. A still wider pulse.

Hammer Hardness Versus Coherence

It does not take much intuition to realize that a hard hammer tip will cause a short duration pulse. This is because the deflection of the hammer tip during the time of the blow is minimal. Since the hammer tip stops rather abruptly, the value of T describing the pulse shape is small.

Thus, a hard-tipped hammer will, because of its short time duration, have its energy spread out over a wide frequency range. The energy at any one particular frequency will be low.

Conversely, a soft-tipped hammer causes a long duration pulse. Most of the energy of such a blow will be in the low-frequency

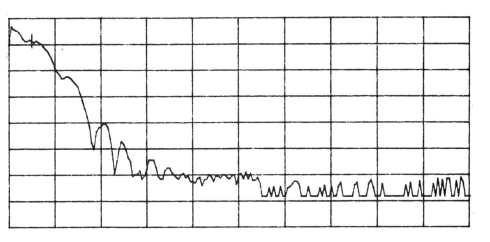

```
                                          AVG=  8
PWR SPECT A      :2.02E+00R              500. HZ      N:   8  F:50HZ
SPAN: 0.000000HZ-10.0000KHZ  SN: -23dBV   FS:  15.00dBR  10dB/
```

Figure B.4. The power spectrum of a 1-lb PCB force hammer with a medium hardness tip.

range of the first $1/T$ lobe. Figure B.4 shows the actual power spectra of a 1-lb PCB force hammer with a medium hardness tip.

Because the energy distribution of the forcing spectra (hammer blow) changes with hammer tip hardness, careful consideration should be given to using the proper tip for the job at hand to insure a good test coherence.

Large, Flexible Machine Suppose that one is testing a large, flexible machine to determine its natural frequencies. The use of a hard-tipped hammer, which approximates an impulse in the time domain, would have a larger proportion of its energy at the high end of the frequency spectra. There will not be much energy available to excite the large mass of the machine with sufficient force to cause it to resonate at the low natural frequencies one would expect to find in a large, flexible machine. The accelerometer motion fed into channel B of the spectrum analyzer, therefore, would be too small to cause a high level of coherence. The test would be inclusive.

The solution to this problem is to use a soft-tip hammer. The resulting wide pulse will cause a large percentage of the energy of the hammer blow to be located in the low frequency region of interest. Thus, barring other test errors, a high level of coherence can be expected, yielding reliable results. This is why a 1-lb force hammer is often adequate to test objects as large as a locomotive.

Light, Stiff Machine In testing a light, stiff structure, the expectation is that the natural frequencies will be high. Although a soft-tip hammer will certainly give the impression of moving the structure, the fact is that there may not be adequate energy at a high enough frequency to resonate the structure at the frequencies of interest.

The coherence of this kind of test can be improved by using a hardtipped hammer. Although the energy of the hammer blow is distributed over a larger frequency range, there will be at least some energy in the high-frequency range of interest.

Cavitation and the Pulse

Cavitation occurs in the flow of a fluid when the pressure in a flow stream drops below the vapor pressure of the liquid, causing vapor bubbles to form. When local pressure further downstream increases above the liquid's vapor pressure, the bubbles implode. Considerable damage results to the metal surfaces in the vicinity of the millions of imploding bubbles. This is shown schematically in Figure B.5.

There are three common causes of vapor formation in a liquid:

1. Flow separation of a viscous fluid from its guiding surface due to a surface discontinuity
2. The addition of heat to the fluid, raising its vapor pressure (boiling point)
3. Reducing the pressure of the fluid to below its vapor pressure (as in the case of too high a vacuum in a liquid-ring vacuum pump)

If a flow is such that small vapor bubbles form and almost immediately implode, the bubbles that collapse will have been small because of inadequate time to grow. The resulting pulses will be quite short in the time domain, approximating white noise in the frequency domain. Thus, cavitation in a small, high-speed pump will sound like a hiss at most and may well be undetectable. The addition to a normal pump spectra of white noise will "fill in the valleys" between the usual peaks of running speed, blade frequency, and so on.

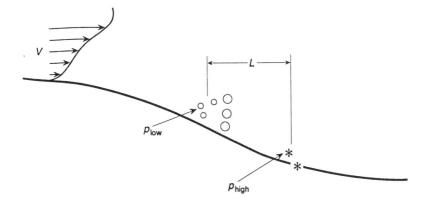

Figure B.5. Cavitation.

Small bubble cavitation may well cause high levels in a shock pulse or spike energy type of bearing fault detector, because the device cannot tell the difference between the white noise of a bad bearing and the white noise of a small bubble implosion. There is no difference. White noise is white noise. More will be said of white noise bearing fault detectors in a later section of this Appendix.

If the above mentioned small bubbles are permitted to grow in size, the pulses caused at their implosion will be wider in time duration. Since the wider pulses will have more energy in the low-frequency range, the modification to the spectra of a normally operating pump will be more dramatically changed by large bubble cavitation.

Bubbles can be allowed to grow by adding heat, dropping the pressure, using a lower vapor pressure liquid, or simply allowing the bubble to spend more time below its vapor pressure as it moves to its high pressure implosion location. Thus, while small, high-speed pumps hiss when cavitating, large, low-speed pumps sound like rocks hitting. Middle-size, medium-speed pumps sound like pebbles. These sounds can be changed by adding heat or pulling a localized vacuum on the liquid. Also, gasoline pumps are likely to have a deeper cavitation sound than a cold-water pump because of the difference in the vapor pressure of these two liquids.

As a result of the way in which the pulse shape of the cavitation of a particular sized bubble affects the vibration spectrum of the pump in question, it is not obvious that cavitation will always show up in as a clear pattern when performing condition monitoring on that pump. If large bubbles are expected, one would look for an increase in broad-band noise in the low-frequency range of the vibration spectra. If small bubbles are expected, condition monitoring will require an examination of the amplitude of the valleys between the peaks.

Rolling-Element Bearing Checkers

There are two basic methods of checking rolling-element bearing condition that have had some success. The classical method is to calculate the frequencies expected in a spectra for each possible component fault of a given bearing (based on the specific geometry of the bearing). We then look for these frequencies in the machine's vibration spectra. If there are no other signals obscuring these low-energy bearing frequencies, a failing bearing can be identified.

The shock pulse and spike energy methods of checking bearing condition are quite similar to each other in their basic operation. Both methods rely on the assumption that a small fault in one of the components of a rolling-element bearing will cause a short duration pulse when it comes in contact with another part of the bearing (for example, an outer race fault contacting a ball).

Since the components of a rolling-element bearing are quite hard, the pulse mentioned above will approximate an impulse. Thus, the signal one may expect from a bearing fault will be white noise in the frequency domain. Remembering that white noise contains all frequencies at equal amplitude, the manufacturers of bearing-checking devices need only follow a routine to observe bearing condition in a frequency region above that normally contaminated by such signals as blade frequency, gear mesh, sidebands, and so on.

A rolling-element bearing checker is basically a meter that reads the output of a band pass filter selected to remove everything except the response of the natural frequency of an accelerometer to a pulse. The operation of such a device can be simulated as follows:

1. Mount an accelerometer with a known natural frequency, usually between 30 and 50 kHz, as close to the bearing as possible. There should be no electronic dampening of the natural frequency.

2. Since the accelerometer has a natural frequency of, say, 36 kHz, it will amplify, by a factor of several orders of magnitude, the level of the signal at 36 kHz. It will not amplify other signals, such as a 400-Hz gear mesh.

3. Put the output signal of the accelerometer through a bandpass filter with a center frequency equal to the accelerometer's natural frequency, in our example, 36 kHz.
4. Measure the amplitude of the output of the bandpass filter.
5. As more faults occur in the bearing, the level of the white noise increases. This increases the amplitude of the output of the bandpass filter, showing incipient failure of the bearing assembly.
6. Older bearing-checking devices simply measured the output of the bandpass filter. One was expected to trend this value in order to trend bearing condition. Newer devices follow the time variation of the output of the filter, crunch some numbers in various algorithms, and display a calculated value that represents the time variation of the white-noise signal caused by a bearing fault.

There are two possible problems in using the shock-pulse method of monitoring bearing condition. One is the inability to differentiate between various sources of white noise. The other is the phenomena of the "healing bearing."

White-Noise Differentiation As stated above, the shock-pulse method cannot differentiate between one source of white noise and another. A bearing checker designed to measure the amplitude of white noise amplified by the natural frequency of its accelerometer may well yield high values due to a bearing fault. The problem, however, is that it will also yield high values due to small bubble cavitation, steam leaks near the accelerometer, turbulence, and so on. The presence of steam leaks near a bearing to be monitored is a common problem in the dryer section of paper mills, where this method, if blindly followed, will lead to the replacement of a great many perfectly good bearings.

Healing Bearings It has occasionally been found by people who regularly take white-noise-type bearing readings that a given bearing will show signs of deteriorating but, shortly before total failure, exhibit lower amplitude values. Machinery monitoring personnel have been fooled into believing that the bearing was getting better by lower

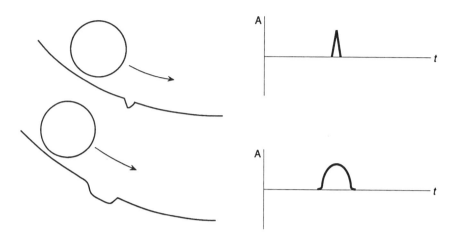

Figure B.6. A healing bearing.

readings when it was actually on the verge of failure. A possible explanation of this phenomenum can be found in pulse theory.

Suppose the bearing was failing not simply by adding more tiny faults (which would increase the white noise content), but because the existing fault was growing. As the fault grew, the time duration of the pulse generated when the fault came in contact with other components of the bearing would grow. The pulse would lose more and more of its white noise characteristics until there was no more energy in the frequency range necessary to excite the natural frequency of the accelerometer. The amplitude reading of the bearing-checking box would drop, implying that the bearing had healed (see Figure B.6).

The above described possibility for a bearing to appear to heal puts an additional requirement on the work of machinery-monitoring personnel. When the reading of a bearing-checking device begins to drop for a particular bearing, careful consideration must be given to the possibility that the bearing is very near to failure. It is wise, at this point, to look at the amplitudes of the classical discrete frequencies of that bearing in order to make a decision as to when it must be changed.

The Walls of Jericho

To demonstrate that pulse theory may be applied to a great many varied phenomena, one last case will be covered: the falling of the walls of Jericho. Jericho is one of the oldest known cities on earth.

The people of Jericho were not very good wall builders. In fact, the walls of Jericho had fallen many times before the time of Joshua.

As can be seen by the background of the photo of some of the remains of the walls of Jericho in Figure B.7, Jericho is located on a fertile plain, not on sand and gravel, as many people think. The area was probably the same in Joshua's time; it was not called "The Land of Milk and Honey" because it was an arid desert. Note that wet, fertile ground transmits low-frequency signals rather well.

Stone walls, especially of the primitive kind shown in the photo, tend to have a large mass and a very little stiffness. Therefore, the natural frequency of the structure would be quite low. Joshua commanded his people to march around the city seven times, blowing ram's horns and shouting and the walls came tumbling down.

Pulse theory yields an interesting insight into this historical event. The blowing ram's horns and shouting would cause pressure waves of rather high frequency, several hundred to several thousand Hertz. It is extremely unlikely that these pressure waves, while certainly capable of scaring the wits out of the citizens of Jericho, would have been capable of causing the low natural frequency walls to collapse.

On the other hand, the stomping of a large number of feet on fertile ground would have caused long-time-duration pulses of rather high amplitude. Pulse theory tells us that long-time-duration pulses have most of their energy concentrated in the low-frequency range, near the natural frequency of the wall.

Figure B.7. The walls of Jericho.

Appendix B

We can surmise, therefore, that the walls of Jericho fell due to the marching of the Jews, not their shouting or blowing or ram's horns. The miracle was not that the walls fell, but that pulse theory was successfully employed by an ancient general at an ancient city to change history.

Something to Think About

The effect of the shape of a pulse is something that is so well known and so widely taken for granted by many of us that its ramifications in vibration analysis are seldom considered. This is unfortunate because, as shown above, the due consideration of the relationship of pulse shape to frequency response can lead to a great many insights and hypotheses that may well provide a better understanding of the phenomena being studied.

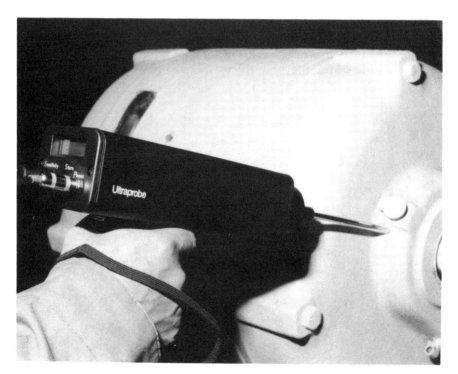

Figure B.8. The Ultraprobe 9000 Bearing Checker by UE Systems.

APPENDIX C

Torsional Vibration

Introduction

Anyone designing a piece of heavy machinery is likely to perform both a lateral (bending) and a torsional natural-frequency analysis of the potential design. This is done to verify that the machine will operate properly in its design range without resonating to destruction. Such calculations are particularly important in systems involving reciprocating equipment and/or a number of flexible couplings or gears.

Although both the lateral- and torsional-vibration characteristics of a machine train are considered to be important in the design stages, hardly anyone ever takes the trouble to actually measure the torsional vibration of the installed machine for either acceptance determination or as an ongoing predictive-maintenance activity. Lateral-vibration-monitoring programs, on the other hand, are extremely popular.

The reason for this obvious inconsistancy between the efforts to measure lateral versus torsional vibration is quite simple. Modern transducers make it easy to measure lateral vibration at each bearing housing of each machine in question. Measuring torsional vibration is more difficult.

In this appendix we will briefly cover the basic mathematics of torsional vibration and then discuss some potential important applications of torsional-vibration measurements. Finally, the more complicated problem of the instrumentation necessary to measure torsional vibration will then be covered.

The Basics

There is a direct analogy between lateral- and torsional-vibration analysis. This can be seen by comparing the two equations of motion below:

Lateral Equation:

$$F = ma$$

where F is the motive force applied to the system, m is the mass (or resistance to motion), and a is the acceleration (motion) of the mass.

Torsional Equation:

$$T = J\alpha$$

where T is the motive torque and J is the mass moment of inertia (or resistance to rotation), and α is the angular acceleration of the rotating component.

Note that the mass moment of inertia is equal to the polar moment of inertia multiplied by its mass. Thus, for a rotating disk, $J = mr^2 = (W/g)(D^2/8)$, where W is the weight of the disk, r is the radius of gyration of the disk, and D is its diameter. Typical units of J would be in inch-pound-second².

With this analogy in mind, we can go from the lateral equation of motion

$$F = m\frac{d^2x}{dt^2} + c\frac{dx}{dt} + kx$$

where x is the displacement of a mass m attached to a viscous damper c and a spring of stiffness k, to the torsional equation of motion

$$T = J\frac{d^2\phi}{dt^2} + b\frac{d\phi}{dt} + k_t x$$

where ϕ is the angular displacement of a rotating disk of J polar moment of inertia, b is the torsional damping, and k_t is the torsional spring stiffness. Note that the units of ϕ are in radians (2π radians = 360°), the units of k_t are in inch-pounds/radian. The units of b are in inch-pound-seconds/radian.

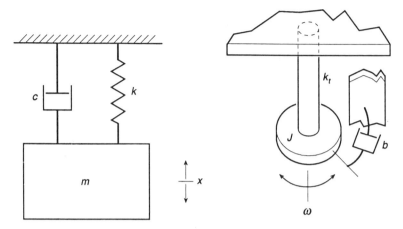

Figure C.1. A comparison of a lateral and a torsional system.

A comparison of the above equations, along with an examination of the two systems in Figure C.1, should be adequate to convince the reader that his level of competence in dealing with torsional problems should be no less than his level of competence in dealing with lateral-vibration problems.

The Advantage of Torsional-Vibration Readings

The information one can obtain from lateral-vibration readings concerning the condition of a machine is well known. To justify the great effort and expense involved in gathering torsional vibration data, one must have a feel for the kinds of unique things than can be learned about a machine from this data.

Reciprocating Machinery

It can now be revealed that the example given in Chapter 1 involving a piston-crank mechanism was wrong. All of the conclusions concerning acceleration, velocity, and displacement were correct, but the proper mechanism for use in the illustrations should have been a slider-crank mechanism.

If one writes the equations of motion of a piston-crank mechanism, it can be seen that one arrives at an infinite Fourier series with all of the harmonic terms present. This means that if one did an FFT spectrum analysis of the rectilinear vibration of a reciprocating machine, all the harmonics of running speed would be present. Thus, a peak in the data at, say, twice running speed, might not indicate misalignment, as it would for a rotating machine. Instead, it would automatically appear simply because the piston went up and down.

Because of this problem, very little rectilinear vibration work has been done on reciprocating machinery. The results are seldom understood.

Torsional vibration, on the other hand, seems to be well suited to reciprocating equipment, because it concerns itself only with the rotational motion that results from the reciprocation of the pistons. Thus, one would look at the nonuniformity of the speed of rotation of the crankshaft (its angular acceleration and deceleration) in order to detect any anomolies in the driving motions. This technique has had some success in the measurement of torsional vibration to detect incipient cracks in vibration absorbers on diesel locomotive engines.

With the aid of synchronous time averaging (discussed in Chapter 2), one might well be able to relate the thermodynamic and fluid dynamic effects at a particular cylinder to its contribution to the rotation of the crankshaft. Thus, a sticking valve, a clogged fuel injector, or an ignition problem in a particular cylinder, which results in a loss of power to the crankshaft from that piston would be seen as a deceleration of the shaft during the time of the power stroke of the affected cylinder. A sticking exhaust valve would have a different timing and characteristic than a sticking inlet valve. A clogged injector or improper ignition would show a larger deceleration of the crankshaft than a leaking valve.

Furthermore, through the use of dual-channel transfer function analysis on a nonoperating machine, one could locate potential torsional resonance problems. This would allow the diagnostician to separate out natural frequency problems from problems in the forcing mechanisms of the system.

Sidebands

Sidebands appear in lateral-vibration spectra as the result of an amplitude or phase modulation of the basic motion of the components being tested. When measuring lateral motion resulting from a rotating forcing mechanism, therefore, it is a secondary effect. If we were to measure the modulation of the rotation directly, via torsional-vibration measurements, these small rotations would be seen as the primary effect. Small changes in the angular acceleration of the shaft would be more prominent in the spectra and the physics of the cause would likely be more obvious than searching the sideband signature characteristics for meaning.

Flexible Couplings

It is well known that a misaligned coupling usually causes high lateral-vibration levels at twice running speed. The highest amplitude of vibration is often in the axial direction, and, if one wished further corroborative evidence of a misalignment, one could check for an approximate 180° phase shift across the coupling.

Although the above lateral-vibration signature will certainly warn the rotating equipment specialist of a shaft misalignment, it may not yield information as to the condition of the coupling. Thus, on a critical machine, one would have no idea of whether the alignment could wait for a scheduled shut down for repair or whether the coupling was about to fly apart.

Sometimes, lateral-vibration analysis can show imminent coupling failure through the existence of a low-energy spike at a frequency equal to the number of, for example, teeth (of a gear coupling) or grids (of a grid coupling) times shaft speed. An observant analyst might infer the condition of a coupling by observation of the sidebands present in the lateral-vibration spectra at the bearing housings on either side of the coupling.

Unfortunately, since a coupling is a very low-energy device, one cannot rely on being able to see either of the two above-mentioned phenomena indicating coupling problems above the ambient noise level of the machine's lateral vibration. One possible solution to this problem is to measure the torsional-vibration characteristics at either side of the coupling. As the transfer function, (for dual-channel analyzers, or the transmissability, for single channel analyzers (a less desirable property to monitor) changes over the course of time, it is an indication that the driven side of the coupling is not following the motion of the driving coupling as it should. This is an indication of coupling wear.

Gears

Gears generate a large number of possible sidebands about the mesh frequency. Sidebands at the mesh frequency plus or minus multiples of the pinion speed and sidebands at the mesh frequency plus or minus multiples of the bull gear speed are typically due to such causes as eccentric gears, a cracked tooth, or nonparallel shafts that allow rotation of one gear to modulate the speed of the other gear. Since the motion that occurs is a combination of both amplitude modulation and frequency modulation, the resulting pattern of sidebands about a mesh frequency is typically a set of asymmetrical peaks at multiples of the running speed of both of the gears.

In the lateral frequency domain, it is necessary to look at the gear signals using the frequency expansion of a spectrum analyzer. If this is not done, it will be impossible to separate the various sideband components from the mesh frequency.

As pointed out in the section on sidebands, the various modulations of rotating motion appear as base-band frequency spikes at the modulating, frequencies. Thus, one can do an analysis of a gear train problem using torsional analysis without the necessity of frequency expansion. This has the advantage that the investigator has the capability of viewing an intuitively illuminating time domain display (angular acceleration versus time). Furthermore, one could view a synchronously time-averaged frequency-domain display to determine whether the observed phenomena is synchronous to one of the gears. A synchronously time-averaged time display could be used to locate the exact position of a particular fault on one of the gears.

Low-Speed Rotation

The detection of the lateral-vibration characteristics of a rotor relies on there being enough energy in the motion of interest to transmit through the bearings supporting the rotor, through the bearing housing, to the transducer (unless, of course, the transducer is a proximeter measuring shaft motion directly). Low-speed rotors, particularly those using oil film bearings, often lack sufficient detectable energy at the bearing housings. Thus, a low-speed roll, such as a coating roll in a paper machine, might not generate sufficient energy to vibration-monitoring devices, while still causing nonuniform density of the product being made.

One proven way to diagnose or monitor these slow-moving rolls is by measuring their torsional vibration. Any problem in the roll drive, such as coupling looseness, worn bearings, and gearing problems, will tend to show up as a nonuniformity in the rotational speed of the roll. This nonuniformity will present itself as a peak in the angular acceleration-frequency spectra.

Once the frequency content of the angular acceleration spectra has been measured, the source can be found. This involves either correlating the disturbances to known forcing frequency sources, looking for coherence between the disturbance and possible sources (this requires a cross-channel spectrum analyzer), or testing for natural frequencies (both lateral and torsional).

Torsional-Vibration Transducers

There are several types of torsional transducers available with adequate frequency response for use with a narrow-band spectrum analyzer. They usually have two things is common: They are both more expensive and more difficult to use than accelerometers. Several of these transducers will be discussed.

Strain Gauges

To measure torsional vibration at any point on a rotating shaft, it is necessary to use strain gauges. These devices are made of small wires of known electrical resistive value. Several strain gauges are bonded to the shaft of interest in a configuration designed to measure the property desired (such as tensile or compressive strain, or torsional strain). As the shaft being measured undergoes local strain in the area of the strain gauges, the strain-gauge wire stretches, changing its resistance. The geometry of the mounting of the gauges, coupled with the way in which each of the resistance changes in the array are compared, indicates the amount of a particular strain in the shaft motion.

The use of strain gauges has typically been used to study the static stresses and strains of a given component under various loadings. Since the frequency response of a strain gauge is limited only by the electronics used to make the measurement and the size of the strain that can be measured (remember, displacement gets smaller as frequency increases), the output of the strain-gauge system can be input to a spectrum analyzer to yield a local torsional displacement versus frequency spectra.

The problems of using strain gauges to monitor the torsional vibration of a shaft are many.

1. The machine must be disassembled to gain access to the location at which the strain gauges are to be mounted.

2. It may be necessary to compensate for the temperature of the shaft by adding an additional strain gauge. Remember that temperature plays a large part in the resistance of a strain-gauge wire.

3. The major problem in using strain gauges on rotating shafts lies in getting the signal off the shaft. Slip rings are cumbersome, difficult to mount on the shaft, and are likely sources of noise in the frequency spectra. Small

frequency-modulated transmitters have successfully been used to transmit the data from the shaft to a receiver connected to a spectrum analyzer. This solution involves the expenditure of approximately $2,000 in signal transmission equipment.

Gear/Sensor Pulse Demodulation

Another way to measure torsional vibration is to mount a precision gear at some location of interest on the shaft and use a stationary magnetic pickup to detect the pulses generated each time a gear tooth passes the pickup. The frequency of this event is obviously the number of teeth on the shaft times the shaft speed. Since the shaft is vibrating torsionally, a set of sidebands about the gear-tooth passing frequency will be generated. The spacings of the sidebands will be proportional to the amount of torsional vibration at the gear. This signal can be demodulated, the carrier frequency (the tooth pass frequency) removed, and the result fed to a narrow-band spectrum analyzer.

Again, there are several problems involved in this method:

1. A major problem is finding a suitable location for mounting the gear on the shaft and the sensor on a rigid, stationary support. Thus, one loses the freedom to observe the torsional vibration anywhere on the shaft.

2. The machining and disassembly/reassembly costs are likely to be prohibitive.

3. The frequency response is limited by the tooth pass frequency.

4. Errors due to gear-tooth error will arise.

Optical Transducers

As an improvement in the gear-based system discussed above, several optical systems exist. Most of these systems use a special tape, which is wrapped around the shaft at a location of interest. A light source (in recent times, a laser) is reflected off the reflective parts of the tape in a manner analogous to the pulses obtained from the gear above. The resulting optical-pulse train is converted into an electrical-pulse train, which is then demodulated.

Since there can be many more light and dark bands on a piece of tape than teeth on a gear, the usable frequency range of an optical device can be significantly higher than the gear system. Some optical systems employ a windowing technique to electronically remove the error at the ends of the tape.

The advantage of using a laser lies in improved readability of the tape in various light conditions. The major disadvantage to the optical system is cost. The typical transducer system is likely to cost approximately $3,000.

Shaft Position Encoders

There are a number of devices on the market that generate a large number of electronic TTL pulses per revolution of the shaft. If one of these devices is mounted at the end of the shaft to be measured, two kinds of signals will be supplied to the spectrum analyzer:

1. A carrier signal at a frequency equal to the number of pulses per revolution times the shaft speed

2. A series of sidebands equally spaced on each side of the carrier signal; the spacings will equal the frequencies present in the torsional vibration of the shaft

A drawback to the use of these devices lies in calibrating the amplitude of the sidebands to correspond to the amplitudes of the torsional motion. The advantage of these devices is that they inexpensively supply the frequency content of the motion. Obviously, the use of position encoders requires that the shaft ends be accessible.

The Hoodwin Torsional Accelerometer

The Hoodwin torsional accelerometer employs a solid metal disk that rotates inside a U-shaped coil in the same way that a disk brake rotates inside the brake caliper. A carrier signal is supplied to the coil. The output of the coil is modulated by the angular acceleration of the rotating disk. This can be input to a spectrum analyzer.

One version of the Hoodwin device is coupled to the end of the shaft of interest. Provision must be made to prevent the coil part of the device from moving. Another version allows the user to machine a disk anywhere on the shaft of interest and mount the coil around it.

The frequency response is superior to that of the geared system or the optical-tape method. The cost of a Hoodwin device is often an order of magnitude less than other methods.

Summary

As can be seen, torsional vibration can be thought about with much less effort than is required for measurement. It is often worth the effort to bite the bullet and install a torsional rotation pickup on a machine of special importance.

APPENDIX D

Condition Monitoring of Reciprocating Equipment

Introduction

In the past 20 years or so, American industry has made great strides toward reducing the time and maintenance costs of rotating machinery, such as pumps, compressors, and turbines. Most of these gains have resulted from the move toward advanced vibration-monitoring techniques given in this book and a philosophy of predictive rather than preventative maintenance. Over this same period of time, reciprocating equipment such as diesel engines and high-pressure compressors have been largely ignored.

The same maintenance staff that regularly stores FFT vibration spectra of its rotating equipment on computer disks for automated monitoring and trending is perfectly satisfied to do no more than change the oil and periodically rebuild their reciprocating equipment without regard to its condition. This wasteful effort is assumed to be a part of industrial life.

There are two likely directions to be taken in dealing with the problem of reciprocating machinery. One of them, torsional analysis, was discussed in Appendix C. This appendix discusses a technique that is somewhat analogous to FFT spectrum analysis in the rotating equipment world, but uses pressure rather than vibration as the machinery health indicator. The instrumentation required has become available only in recent years, but some of the theory has been available since the early 18th century.

Why Not Use Frequency Domain Data?

The technique of predictive maintenance for rotating machinery begins with the gathering of vibration data taken at the bearings of the machine in question. This data is of the form of amplitude (in in/sec or g) over time. The resulting signal can be viewed on a simple oscilloscope. Unfortunately, the time trace on the oscilloscope is made up of a combination of so many different sine waves that it is almost impossible to make sense of it without spending vast amounts of time and effort.

The problem is greatly simplified if the complex time waveform is transformed into the frequency domain via an FFT device such as a so-called real-time spectrum analyzer. The resulting display of amplitude versus frequency, or spectra, can be easily examined for such things as machine running speed to check balance, two and three times running speed for signs of misalignment, and so on.

The problem of the reciprocating machine, as mentioned in the section on torsional analysis, lies in a single fact: If one writes the equations of motion of a piston-crank mechanism, it can be seen that one arrives at an infinite Fourier series with all of the harmonic terms present. This means that if one did an FFT of the rectilinear vibration of a reciprocating machine, all harmonics of running speed would be present. Thus, a peak in the data at, say, twice running speed, might not indicate misalignment, as it would for a rotating machine. It would automatically appear simply because the piston went up and down.

Because of this problem, very little rectilinear vibration work has been done on reciprocating machinery. People seldom understand the results. A second technique used more in the development of new equipment is called acoustic intensity. This method (described in Chapter 9) uses a pair of microphones to generate, through cross-spectra calculations, a sort of acoustic vector capable of pointing at noise sources located deep in the bowels of the machine. The instrumentation required for acoustic intensity is typically far too elaborate and time consuming for routine monitoring.

Pressure as a Machinery Health Indicator

Monitoring the condition of reciprocating equipment was not always a problem. A century ago, when reciprocating machinery ran at 100 rpm, people used P-V diagrams to determine what was thermodynamically happening inside the cylinders. A paper and pencil arrangement was attached to the machine in question such that the pencil moved vertically according to the pressure in the cylinder and horizontally according to the position in the stroke. The resultant *P-V* diagram could be compared to the results one would expect from a thermodynamically ideal machine (e.g., a Carnot, engine) of the same physical dimensions and operating conditions. The problem with this ancient but insightful method is that it would be difficult to avoid flying pencils at shaft speeds of 1,000 rpm.

Recent advances in transducer capability and the advent of capable digital oscilloscopes now allow us to replicate the pencil drawn *P-V* diagrams of old on any reciprocating machine at any speed. Furthermore, it is now possible to generate the equivalent of a *P-V* diagram for any number of manifolded cylinders under time-varying loads with a single measurement through the method of synchronous time averaging as described in Chapter 2 for FFT spectrum analyzers.

In synchronous time averaging, the input buffer of the digital oscilloscope begins taking its time data when it receives an external trigger signal from the rotating shaft. After the last datum required by the buffer for single average, the oscilloscope stops and waits for another external pulse before reloading the buffer.

If the external pulse comes from an optical or magnetic pick-up looking at a keyway on a shaft, the time data gathered will always start at the same shaft position. If the time signal is averaged over a large number of data blocks, therefore, all events that are related to shaft position will reinforce themselves. Those signals that are asynchronous with shaft motion will appear to be random in nature and eventually average out to a low value. Since we know the relationship between the reference pulse and the shaft position (by having kept track of the geometric relationship between the keyway or reflective tape and the pick-up position), we are able to relate any event noticed in the amplitude-over-time data to a particular crank position and, hence, to a particular cylinder.

Generating the *P-V* Diagram

Employing the notion of synchronous time averaging, the analysis of the thermodynamic condition of the reciprocating machine using *P-V* diagrams for each cylinder is straightforward. We need only two things: (1) Obtain a once-per-revolution signal from the shaft of interest (if an optical pickup is used, this requires only the placement of a piece of reflective tape on the shaft) for the external trigger jack of the digital oscilloscope. (2) Place a piezoelectric pressure transducer judiciously in the exhaust manifold of the machine to be tested.

The sampling duration of the oscilloscope must be long enough to include at least one revolution of the shaft. It is better, however, to include a great many revolutions of the shaft in order to see time variations in the signal. This requires an oscilloscope capable of storing a large amount of digitized data for later analysis. A 32-kbyte memory capability and the superior amplitude resolution of the time domain display of a properly selected digital oscilloscope makes possible drawing valid conclusions about the subject machine. Trying to get good time data from the time trace of an FFT spectrum analyzer would be difficult since it usually divides the memory period into only 1,024 segments.

The Tests

The following section describes the test setup and the results obtained from instrumenting a diesel engine (which had been lying dormant for 6 months) in the way suggested above. The data presented in this section was taken on a 4-cylinder Caterpillar diesel-dynomometer at the New York State Maritime College at Fort Schuyler in New York. It represents the kind of results that can be obtained by obtaining synchronously time-averaged exhaust pressure data using a digital oscilloscope with the capability of very high sample rates and a large memory.

Setup

To determine the practicality of the above approach, a 4-cylinder Caterpillar diesel driving a dynomometer at the New York State Maritime College at Fort Schuyler, New York was used.

The pressure transducer was mounted on the end of a short copper tube attached to the exhaust manifold. Although a more ideal location could easily be found on the exhaust manifold, the tubing was used to avoid damaging the transducer's built-in amplifier circuit with hot exhaust gasses. A piece of reflective tape was placed on the output shaft of the dynomometer for use as a trigger.

The digital oscilloscope was set up to oversample the data to insure good resolution and nonaliased data. A large memory buffer was allocated to save a great many synchronously averaged pressure pulses.

Results

After the engine was warmed up, pressure data was taken at the exhaust manifold at 2,000 rpm and no load. Fifty six synchronous time averages were performed. From the plot of Figure D.1 the following can be seen:

The lower trace shows a regularly repeating pattern for about 18 shaft revolutions following the trigger event, as well as two revolutions before the trigger set point (it is often useful to set up the oscilloscope to display some pretrigger data to see what events were taking place before the trigger corrected). Note that these 20 shaft revolutions actually represent 56 synchronous time-averaged sets of 20 revolutions each. One advantage of having 32 kbytes of memory in the oscilloscope is the ability to see what is happening over relatively long time spans with adequate resolution. In this case, the long time history shows that the engine is running in a steady-state condition.

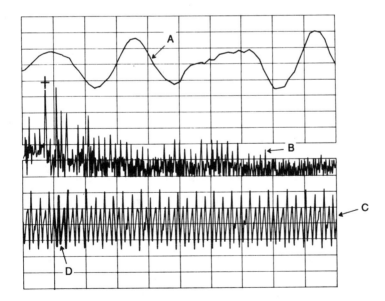

Figure D.1.   Caterpillar Diesel. 2/5/87, 2,000 rpm (no load), exhaust pressure of four synchronously time-averaged cycles. A. Expansion of the four amplitude vs. time peaks (one engine cycle) indicated by "D." B. Fft of the amplitude vs. time data indicated by "D." C. Amplitude vs. time data for many engine cycles. D. One engine cycle.

The upper trace of Figure D.1 is an expansion of the one complete cycle that occurred at the trigger point. This expansion can be seen in the lower trace as the four pressure pulses that have been intensified beginning with the trigger marker arrow. Note that two of the cylinders are generating less exhaust pressure than the other two. A probable cause of this will be discussed later.

The center trace of Figure D.1 is an FFT (pressure versus frequency plot) of the averaged time history. The cursor marker is at twice crank speed. Note that the signal is rich in harmonic content.

Next, the engine was loaded up to 40% of full load and data was again taken at 2,000 rpm. This can be seen in Figure D.2. Note the following:

- The long pressure-time history in the lower trace is not quite as stable as was found in the unloaded condition. This is to be expected as the engine is now trying to do work against the dynomometer.

- The four pressure pulses taken just after the trigger point show that the output of each of the 4 cylinders is essentially the same. A logical explanation for this

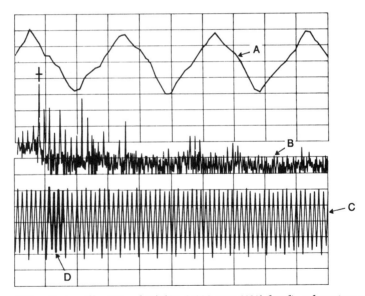

Figure D.2. Caterpillar Diesel. 2/5/87, 2,000 rpm (40% load), exhaust pressure of four synchronously time-averaged cycles. A. Expansion of the four amplitude vs. time peaks (one engine cycle) indicated by "D." B. Fft of the amplitude vs. time data indicated by "D." C. Amplitude vs. time data for many engine cycles. D. One engine cycle.

symmetry as opposed to the asymmetry seen in Figure D.1 is that two of the injectors were partially clogged after 6 months of standing idle and were eventually blown out.

- The FFT data is somewhat different from that found in Figure D.1. There seems to be more energy at some of the low-frequency harmonics and less at high frequencies. The meaning of this is unknown.

The engine was then brought up to 100% load. It was on the verge of stalling. Figure D.3 shows some interesting information, including the following:

- The lower trace, which displays many averaged cycles from the trigger point, shows that there is a low-frequency pressure modulation or surge affecting all cylinders. If it were not for the ability of the oscilloscope to store digitized data far beyond the trigger point, this surging might have been ignored by someone looking at a set of averages of a single revolution after the trigger point.

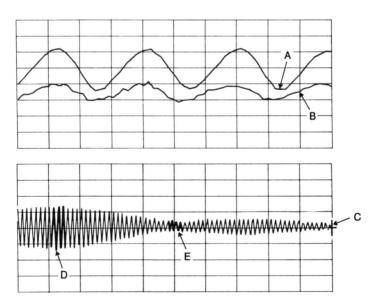

Figure D.3. Caterpillar Diesel. 2/5/87, 2, 000 rpm (100% load), exhaust pressure of four synchronously time-averaged cycles. A. Expansion of the first set of four amplitude vs. time peaks, indicated by "D." B. Expansion of the second set of four amplitude vs. time peaks, indicated by "E." C. Amplitude vs. time data for many engine cycles.

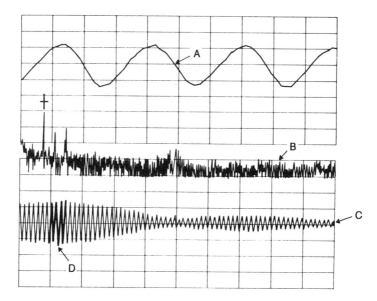

Figure D.4. Caterpillar Diesel. 2/5/87, 2,000 rpm (100% load), exhaust pressure of four synchronously time-averaged cycles. A. Expansion of the four amplitude vs. time peaks (one engine cycle) indicated by "D." B. Fft of the amplitude vs. time data indicated by "D." C. Amplitude vs. time data for many engine cycles. D. One engine cycle.

- The upper two traces show the four pressure pulsations at the trigger point compared to the pulsations at the center of the time axis. Note that they are not only different in amplitude, but the shape of some of the low-amplitude pulses have lost their smooth appearance. This is an incipient stall.
- The FFT of this data is shown in Figure D.4. Note that most of the harmonic content has disappeared into the noise level.

In order to verify our assumption about having blown out whatever was clogging some of the injectors at the beginning of our test, the engine was brought back to no load operating condition and Figure D.5 was plotted. By comparison to Figure D.1, the following can be seen:

- Both the trace of a large number of cycles of the shaft and the trace of the four pulsations at the trigger point are far more symmetrical than during the initial test.
- The FFT of this data has less harmonic content than was the case with the initial no load test.

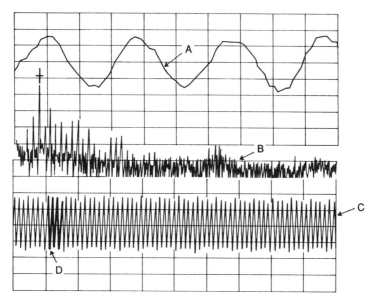

Figure D.5. Caterpillar Diesel. 2/5/87, 2,000 rpm (no load), exhaust pressure of four synchronously time-averaged cycles, 2nd trans. A. Expansion of the four amplitude vs. time peaks (one engine cycle) indicated by "D." B. Fft of the amplitude vs. time data indicated by "D." C. Amplitude vs. time data for many engine cycles. D. One engine cycle.

Summary

The test described above clearly demonstrates how much information can be learned about reciprocating machinery condition by synchronously time averaging the pressure pulsations at the exhaust manifold. This is obviously a far less painful chore than monitoring pressure at each cylinder, however, much work still needs to be done.

It is possible to extract the exhaust pressure pulsations of a particular cylinder of a 4-cylinder diesel engine by mounting a piezoelectric pressure transducer in the manifolded exhaust gas flow and performing a synchronous time average over a great many cycles. Comparison of the pressure pulsations of one of the cylinders to the others can be used to detect anomalies affecting engine performance.

It can be seen from the test data that it is important to be able to store a large number of cycles of rotation after each trigger point in order to get an indication of any relatively long-duration modulating effects occuring in the engine.

The time trace of a number of synchronously averaged pressure pulsations can be seen to be analogous to the ancient *P–V* diagrams done in the days of very-low-speed equipment. As such, these traces give a more intuitively obvious picture of the events taking place in a particular cylinder than can be had from frequency-domain data.

APPENDIX E

Balancing

Introduction

It has been said that 85% of the vibration problems in the world can be solved by balancing. This section has been placed at the back of the book, because, it is hoped, the reader now knows better. A great many other possible forcing frequencies have been described. The possibility should be well understood by now that a high level of vibration can be caused, not by balance or any other forcing mechanism, but by the existence of a natural frequency near what would otherwise be an acceptable level of forcing excitation.

The author has met several vibration consultants who have described a number of fans or pumps in a given facility that must be balanced every 3 months or so. These problems have the following characteristics:

- Most of the vibrational energy is at running speed
- There is no reason to suspect a buildup or erosion on the fan or impeller blades
- The possibility that a natural frequency near running speed has not been checked
- The unhappy owner of the machinery in question pays his bills on time

Of course, most consultants are not disreputable, but some have college degrees in Economics, not Engineering. It will be assumed here that the reader has learned enough to check transfer functions and coherence for signs of a resonance problem and has tried synchronous time averaging to verify that the high-amplitude once-per-revolution signal is, indeed, synchronous with the shaft.

If balancing is necessary, accidental destruction of the machine is not impossible; an improperly mounted trial weight can fly off the machine or can be dropped into it. Some companies are capable of performing a balance job using in-house personnel, but send for an outside consultant because of liability. Balancing requires liability insurance as it is the only "vibration analysis" function that is not nonintrusive.

The most common ways to balance are to use an analog balance machine that has a strobe light for phase or an FFT device in the synchronous time averaging mode triggered off the shaft to be balanced to obtain magnitude and phase. Polar plots have been made obsolete by the many algorithims available for hand-held calculators for single- or multiplane balancing.

The Causes of Unbalance

As pointed out in Chapter 4, unbalance results from the fact that the center of gravity of a rotating member does not coincide with the center of rotation. This causes the creation of a centrifugal force vector (as seen in Figure E.1) pointing radially out from the center of rotation and rotating at a speed equal to (and synchronous with) the speed of the rotating member itself. The amplitude of this force vector is equal to

$$F_{bal} = Me\omega^2$$

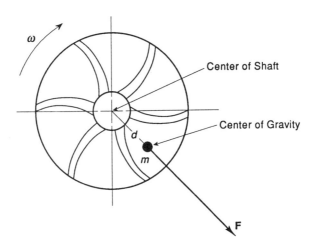

Figure E.1. A rotating unbalanced force vector.

where M is the unbalanced mass, or W/g, e is the distance between the center of rotation and the center of gravity, and ω is the rotational speed in rad/sec, or $2\pi f$ Hz.

With this in mind, one can imagine several likely sources for unbalance:

- Manufacturing problems, such as nonuniform castings, cocked assembly, bent shafts, and nonconcentric machining

- Nonsymetrical reduction of mass of the rotating element from wear, erosion, corrosion, or blade breakage

- Nonsymetrical mass addition, such as product buildup on pump or compressor blades or the ingestion of loose parts or stray animals

- Thermal expansion or contraction sufficient to cause shaft bending or internal misalignment

- The application of an external force large enough to bend the shaft or rotor, for example, condensation of a vapor to a liquid in a vapor handling machine such as an ethelyne compressor in a refinery, ingestion of a slug of water into a centrifugal vacuum pump in a paper mill, attempting to align a pair of machines by pushing on a shaft with a long bar

A further cause of imbalance often occurs when the machine in question runs at or near a natural frequency. In this case, the mode shape of the shaft is such that some part of the rotating element is far enough from the normal center of rotation that an unbalanced centrifugal force appears. This case is quite different from the "rigid body" sources of unbalance mentioned above, and must be handled differently.

Factors to Consider When Balancing

A severely unbalanced machine may not only destroy itself, but can tear itself from its foundation and fly through the air, risking damage to other machinery and people nearby. There is, therefore, a significant economic motivation for correcting a condition of poor balance as soon as possible.

One of the major issues involved in balancing a machine is the production time lost while balancing. Since heavy machinery cannot be turned on and off like a light bulb, the time required to balance is not a function of the calculations to be done. It is almost always dependent on the number of starts per hour allowed on the machine and the time it takes to add trial or corrective weights to the rotating element. Thus, an efficiently done balance job might will require a full day. A trial-and-error method of balancing might take a full week, with poor results.

Some of the major risks of balancing a machine are the possibility of damaging it by an improper disassembly or reassembly or by dropping a balance weight deep into the bowels of the machine. Too large a trial weight in the wrong position could well cause the unbalance to become so large that the machine might be wrecked before it can be brought to a stop.

Before one attempts to balance a large machine, one should use the knowledge learned in Chapter 4 to verify that the problem is, indeed, likely to be one of balance. Remember that a true balance problem is

- Synchronous to the shaft
- Is a function of the square of the speed
- Has a repeatable phase reading relative to a trigger signal derived from a once-per-revolution signal synchronous to the shaft (as from an optical pickup looking at a piece of shaft-mounted reflective tape)

The possibility of a structural problem should also be eliminated via the use of transfer functions and coherence. Balancing should be a justified attempt to correct a diagnosed problem, not the first shot at a quick solution to a complex problem.

The Philosophical Problem with Balancing

Balancing a rotating machine would be child's play if we knew exactly what level of vibration resulted at each bearing due to a unit amount of unbalance in the rotor and if it were true that the phase angle of the vibration due to unbalance corresponded exactly to the position of the actual unbalance in the rotor. If this were the case, it would be necessary only to make a running vibration reading at each bearing and calculate the amount and position of the correction weight. There would be no trial weights or additional machine startups.

Unfortunately, every machine has a unique transfer function (magnitude and phase) between the rotor and bearings. Since this transfer function is usually unknown, it severely clouds the issue of the amount of unbalance and its position on the rotor as inferred by the vibration readings on the bearings alone. One must therefore resort to the game of trying to "back into" the relevant aspects of the system's transfer function by using trial weights.

Thus, the scenario proceeds as follows:

1. An unacceptable level of vibration is found at the running speed of the machine and is verified as a true balance problem

2. The amplitude of the vibration is recorded and its phasal relationship to either an arbitrary mark on the shaft that is illuminated by a strobe or to a trigger reference used for synchronous time averaging is noted

3. A trial weight is added to the rotor and a new amplitude and phase of vibration at the bearing is measured

4. Since we now know that a trial weight of mass M located at a specific position on the rotor caused a certain shift of the original vector due to unbalance by a particular amount (as measured at the bearings), we can now work backward to establish where and how large the trial weight should have been to move the unbalance vector such that it is as small in amplitude as possible

5. The correct weight is added to the rotor and, if the transfer function for the machine remains linear, balance is achieved

6. If the transfer function has changed somewhat by the improved balance, a few iterations of the steps above would be required to bring the balance to acceptable levels

Note that the influence coefficients found as a result of adding the trial weights are really components of the transfer function described in Chapter 5. If these values are saved, and if the machine doesn't change structurally in the future, it should not be necessary to go through the time and effort of using trial weights to balance this particular machine in the future.

Vectors

The easiest way to understand balancing is to look at vectors. While numbers (or scalars) only have an amplitude value, vectors have amplitude and direction. Thus, 30 mph is a scalar amplitude value; 30 mph due south is a vector value. For balancing purposes, vectors are best drawn on polar graph paper. The length of the vector can be shown as proportional to length on the graph and direction by the angle. Figure E.2, for instance, shows a vector of 3 units of amplitude and 30° from some reference point. In this case, we might imagine the 3 units to equal, say 3 g rms.

Vectors are distinguished from scalars by showing them with their directional angles. A vector of amplitude A and an angle of 28° is shown as

$$A \langle 28°$$

Vector addition will be described in order to give a physical feel for the process of balancing. Figure E.3 shows the addition of two vectors $A \langle a = 3 \langle 30°$ and $B \langle b = 5 \langle 15°$.

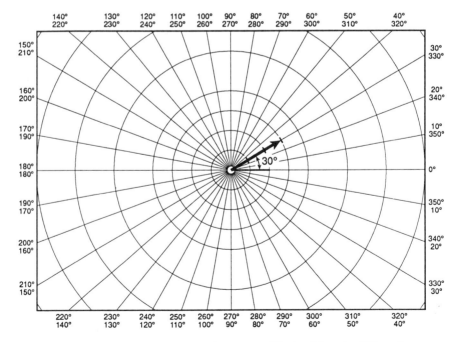

Figure E.2. *A vector representing an amplitude of 3 g at a 30° angle.*

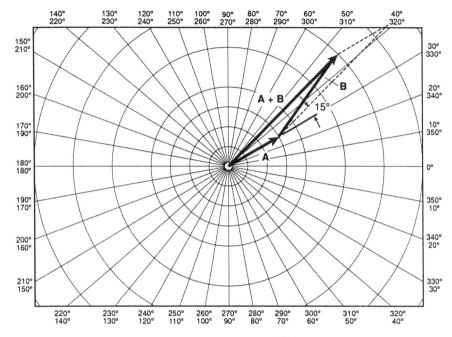

Figure E.3. Vector addition.

Note, in Figure E.3, that one can graphically add vectors by constructing a parallelogram of the two vectors to be added and drawing the result by connecting the initial point of the two vectors to the final point of the parallelogram.

Single-Plane Balancing

With the advent of the sophisticated handheld calculator, it is an almost trivial matter to perform a good balance on a complex rotor assembly. One simply enters the proper algorithm to handle the kind of vector calculations required for the geometry of the system and adds some data measured from a vibration meter (with and without trial weights), the proper amount and location of the balance weight to be added or subtracted then appears on the calculator display. The only problem with this method is that, if the user has no physical feel for what the calculator is doing, he will have no physical feel for what he is doing either. The balance process will then become a new experience each time it is done, with mediocre results and the wasting of time and effort.

A single-plane balance will be discussed in all its laborious detail so that the balancing technician can see what is physically happening. Hopefully, he can then buy a balancing program and never have to do a vector analysis again. Thus, the pain of a vector balance should be endured once in each person's lifetime.

Amplitude and Phase Data

It will be assumed that the balance will be performed using a single-channel spectrum analyzer. To get phase information, the analyzer must synchronously time average. All phase angles will be relative to the arbitrarily selected trigger position on the shaft used to trigger the analyzer. So that small changes in rotor speed due to the addition of trial masses do not affect the readings obtained from the analyzer, the analysis range of the analyzer should be chosen such that the running speed of the machine falls in the lower 20% of the frequency range. This condition is not necessary in those analyzers offering an "improved accuracy" button or when order tracking (causing cell-centered data at running speed).

The phase values obtained in the above setup represent the lag between the time the analyzer begins to take word number one of the 1,024 words that make up a single time average for a 400-line analyzer and the time at which the peak of the sine wave at running speed occurs at the transducer. This is analogous to the phase obtained by illuminating a reference mark of a shaft by use of a strobe light, which fires at the peak of the sign wave seen by a tunable filter balance analyzer.

"As Is" Reading

The first reading obtained is the "as is" reading taken at a bearing of the machine in the radial direction at the speed and operating conditions for which the machine is to be balanced. The resulting vector is drawn from the center of the polar plot radially out a length equal to the measured amplitude of vibration and at the measured phase angle. This is shown as vector **A** in Figure E.4.

Trial Weight Selection

The mass of the trial weight should be large enough to effect the dynamics of the rotating system without being large enough to wreck the machine. Many balancing specialists suggest a trial force equal to about 30% of the weight of the rotor. Thus, by the equation for centrifugal force:

$$\text{mass of trial weight} = 0.3 \times \text{rotor weight} / (R \times \omega^2)$$

Balancing 251

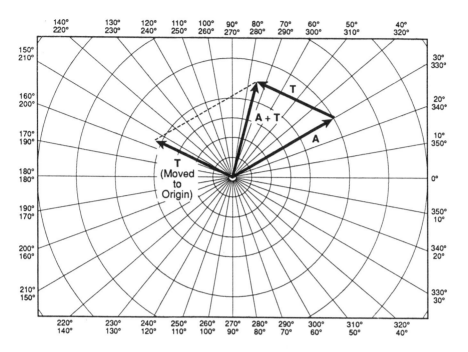

Figure E.4. A vector balance.

where the trial mass and the rotor weight have consistent units, R is the radius at which the trial mass will be mounted, and ω is the running speed of the machine in rad/sec.

Mounting the Trial Weight

Mount the trial weight at a radius R from the center of rotation. A convenient angular location may be estimated by mounting it 180° from the phase reading obtained above. The correctness of this location as a final balancing position depends on the transfer function of the structure.

It is crucial that the trial weight be mounted rigidly and safely to the rotating structure. Remember that centrifugal force will try to break the trial weight free, causing untold damage.

Trial Weight Readings

With the machine brought up to the correct speed, obtain the new values for the amplitude and phase of the vibration at the measuring point. These values represent the vector of vibration caused by the original unbalance plus the trial unbalance. This is shown in Figure E.4 as the vector **A + T**.

Calculate the Influence Coefficients

The vector **T** in Figure E.4 is obtained by constructing a vector from the tip of **A** to the tip of **A** + **T**. This is the vector, moved to the center of the polar plot, that would have existed in the specific subject machine if it had been perfectly balanced before the trial weight was added. We now know, therefore, how the machine responds to a given amount of unbalance. The influence coefficients can be calculated.

Since the trial weight caused the unbalance vector **T**, the amount of correction weight (at the same radius) needed to correct the original unbalance vector **A** will be

$$\text{correction weight} = \text{trial weight} \times A/T$$

in consistent units.

The position of the correction is found by measuring the included angle between **A** and **T** and adding the balance weight that number of degrees opposite to the direction of the phase shift caused by the trial weight. If the angle required by this method cannot be conveniently obtained, the weight can be split into two smaller weights in the same plane mounted at the nearest convenient locations. This is determined by finding the amplitudes of the two weights which, when drawn on polar paper at their mounting angles, have a resultant equal to the correction weight at its calculated angle.

If a closer balance is required, or if the amount of balance weight was enough to cause a small change in the system's transfer function, a second iteration of balance should be performed using the above result as a starting point. After the balance has been completed, it is wise to save the values of the influence coefficient for future reference.

Multiplane Balancing

The single-plane balance problem described above was simple enough to allow one to grasp the fundamental principles of balancing. Unfortunately, most balance problems cannot be solved simply by placing balance weights in a single plane. This is because most machines have the mass of their rotors distributed over their axial length, rather than having a width that is very narrow compared to its diameter. This means that a multiplane balance is required to avoid creating a couple-type unbalance in the rotor.

In a multiplane balance, one must determine the effect that a trial weight at each of the planes has on the rest of the rotating system. Thus, the number of runs required to establish an adequate balance is significantly increased, as is the number of computations involved. Since one can easily purchase handheld-calculator programs to perform multiplane balancing, there is little need to go through a manual example here. Follow the instructions in your handheld-calculator program.

Rigid or Flexible Rotors

A rotating element that does not operate near a critical speed is referred to as a rigid rotor. In such cases, balancing can proceed as described above and weights can be added in fairly obvious and convenient planes. A rigid rotor balanced to a close enough tolerance at one speed will remain adequately balanced at other speeds at which the rotor remains rigid.

A rotor is said to be flexible if it is operating near a critical speed. In this case, the location of balancing planes is not at all obvious, and a machine balanced while operating in the region of one particular critical speed will almost always be out of balance at another speed. This is because the mode shape of the shaft is different at each critical speed and a balance weight added at another speed may well be at a position on the shaft that is deflected quite differently at a particular critical frequency.

Suppose, for instance, one balanced the shaft shown on Figure E.5 in planes A and B at a speed at which the rotor is rigid. At a lateral critical speed of the shaft, the mode shape of the shaft is such that planes A and B are at nodes. The additional weight added in these planes will have almost no effect on the balance of the shaft. The problem planes will be at C, D, and E. Balancing must be done at these planes to reduce vibration.

The above problem shows up often in large steam turbines as well as gas turbines. They typically run above their first critical speed and unbalance tends to be due to some strange mode shape. The mode shape must be calculated using a finite element program on a computer in order to properly balance the machine. Fortunately, the turbine manufacturer usually has this information and can lend significant guidance in solving the problem.

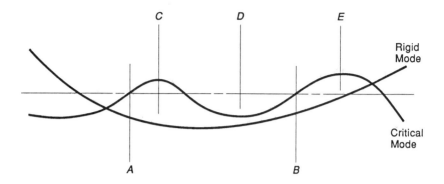

Figure E.5. Rigid versus flexible modes.

Summary

The time-consuming work of polar plotting to balance a machine has been replaced with the pushing of a few keys on a handheld calculator. What has not been replaced with the advent of the calculator is the thinking that must go into deciding whether a vibration problem can be solved by balancing or not. Also, it is important to think about the physics of balancing to be sure that what the calculator says makes sense.

The decision as to whether the rotor is rigid or flexible, as well as the number of planes required for a proper balance, must still be made by the human. Read the calculator balance program manual and think about the physics of what you are doing.

APPENDIX F
Paper Machine Speed-Ups

The decision to speed up a paper machine, whether it produces tissue, kraft, or coated papers, usually results from economic and marketing concerns. Since a paper machine represents an extremely large capital investment, it is natural to try to achieve the maximum output/hour.

After the desired output has been determined, the speed of the machine necessary to deliver that can also be set. Generally, since several grades of paper are made on a specific machine, a desired speed range is prescribed.

Next, decisions on necessary drying capacity, felt and wire configurations, and any new drive requirements are made. A machine modification plan then can be devised and implemented.

One factor often ignored in the planning of a machine speed-up is the resulting change in vibration of the machine's numerous components. This can lead to disastrous results. In the normal course of events, a mill never finds out that the excessive vibration of some particular roll will prevent the machine from achieving planned operating speeds until a commitment for delivery of the paper has been long established and cannot be met.

An example of the need to anticipate these vibration problems came as a result of investigations made on a twin-wire tissue machine at a large paper facility. A speed-up was planned in two stages. The first stage was to be a 7.5% speed increase, and the second stage was to be a 19% increase.

The first stage speed-up was such a failure that more than a year was spent investigating the problem and determining possible solutions. During the investigation a great deal of lost production occurred due to a number of journal failures and the inability to run the machine at the planned speed. Investigation results identified the problem as an excitation of a natural frequency of a press roll caused by the increase in frequency of the three-times-running speed forcing mechanism of the roll. As a result, a method was devised to easily test for potential vibration problems from the anticipated speed-up of three other tissue machines.

Advanced Vibration Techniques

A significant number of people in the paper industry understand the use of single-channel FFT vibration measurements as a predictive maintenance tool. Unfortunately, the solution to the above problem, and the method devised to predict this type of problem for future speed-ups requires the use of a dual-channel FFT analyzer capable of determining the transfer functions of the various structures and the coherence associated with those transfer functions. The following is a brief description of the transfer function and coherence, and an outline of how to obtain them.

Natural frequencies

All forcing frequencies share the characteristic that they are self-generated. Some common examples of these are the once-per-revolution signal caused by unbalance, the twice-per-revolution signal caused by misalignment, and the three-times-running speed spike caused by the blade frequency of a three-bladed pump. These result from the purposeful design and the imperfect manufacture of a particular machine.

If the machine under study is turned on, the forcing frequencies, at whatever level, appear. If the machine is shut off, the forcing frequencies disappear. Most importantly for this article, if the machine changes speed, the forcing frequencies shift proportionally.

Natural frequencies (or resonances) are quite different in character. They are due to the nature of the structure of the machinery, piping, and support system. They are not self-excited, but can be viewed as lurking within the structure of the system, ready to cause violent reactions when excited. They result from the values of the mass, stiffness, and damping of a structure and are not usually a function of the operation of the machine. Therefore, when analyzing the vibration of troublesome machinery, it is often quite important to determine the natural frequencies of the machine-support system.

Structural response of a simple vibrator

A natural frequency is one in which the system exhibits very large magnitudes of vibration when excited by a very small force. Every real structural system has infinitely many natural frequencies. When a system is wildly vibrating at its natural frequency, it is said to be in resonance.

In most cases, it is possible to consider the response of a complex structure as the summation of all its different resonances. The motion at any given natural frequency can be studied without considering the effect of other, widely spaced natural frequencies. Thus, a complex structure can be modeled as a set of simple spring-mass systems.

The curve of Figure F.1B shows the response of a single degree of freedom system. Such a system has the property that its motion can be described by only one parameter, such as the motion along the vertical axis of the spring mass system of Figure F.1A. This has been described as being the last vibratory system to be understood by man.

The curve has, for its vertical axis, the value of the resultant motion/unit of excitation force. The horizontal axis has the value of frequency divided by natural frequency. Thus, if a person is interested in what happens to this system when excited by a force of two lbs at a frequency of three times the natural frequency (note that the value of the natural frequency has not been specified), that person would read the value of motion/force that corresponds to the horizontal value of three and double it (since the force of interest is two units in strength). It is obvious that if a forcing frequency is near the natural frequency, very large levels of vibration can be expected.

Measuring the natural frequency

There are two functions, available only in multichannel FFT spectrum analyzers, that make dual-channel analysis extremely useful for locating natural frequencies—transfer function and coherence. These functions can be explained in nonmathematical, practical terms.

The transfer function can be thought of as a system output (both magnitude and phase) divided by an input (magnitude and phase). The most common transfer function used in machinery analysis is motion (acceleration, velocity, or displacement) divided by excitation force or:

$$TF = motion/force$$

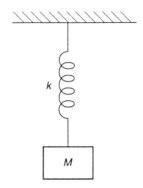

Figure F.1.A   A single degree of freedom system.

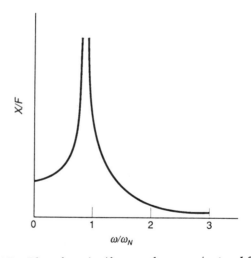

Figure F.1.B   Plot of motion/force vs frequency/natural frequency.

One example of a simple transfer function is the natural frequency plot shown in Figure F.2. Here, motion/force and phase difference are plotted vs frequency.

For linear systems, once the transfer function is obtained using any force (having an adequate frequency content), the motion can be predicted for any other force. That is:

$$\text{New acceleration} = \text{new force} \times \text{TF}$$

Paper Machine Speed-Ups 259

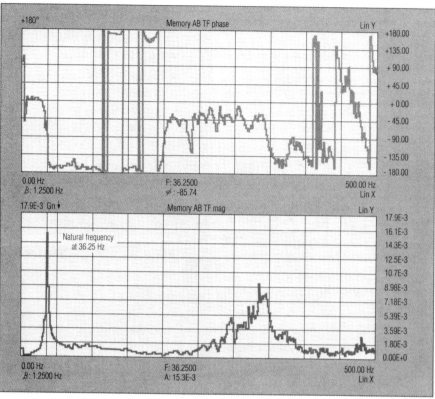

Figure F.2. Transfer function of a centrifugal pump. Top: phase vs frequency. Bottom: acceleration/force vs frequency.

Since it is possible to characterize a linear system by establishing the transfer function using any input force containing the frequencies of interest, a dual-channel spectrum analyzer test for transfer function can be carried out quite simply. In an impact test, a hammer with a force transducer is used as the excitation source (Channel A), and an accelerometer signal is used as the output (Channel B). If the point of impact and the accelerometer mounting location are well chosen, the natural frequencies of the system can be determined easily by locating the peaks of the transfer function magnitude plot or the point of 90° phase shift. Such a test would be conducted as follows:

1. Mount an accelerometer on the machine component in question and connect it to Channel B of a two-channel analyzer.
2. Connect an instrumented force hammer to Channel A.
3. Stop the machine. It is desirable that the only excitation to the machine come from the instrumented hammer blows.

4. Strike the machine a few times with the hammer. These blows will be needed to set up the analyzer for the test. This is often the most tedious part of the entire test.
5. Set the analyzer's trigger level (Channel A). Set the input attenuation of Channels A and B to avoid overload. Change the weighting window of the analyzer from Hanning to Flat. Better analyzers will also ignore any data gathered during an overload condition.
6. Choose a time window (1/bandwidth) that shows the proper ring down of the time domain output of the system.
7. Average several blows. View the time domain signals of both channels during this test to assure yourself that the blows being imparted are good clean blows. Some analyzers allow the use of special weighting windows to help assure this.
8. View the transfer function magnitude and phase. The natural frequencies of the system are those points where the magnitude-vs-frequency plot peaks, and the phase-vs-frequency plot shows a phase shift of 90° for the force gauge/accelerometer instrumentation.
9. View the coherence display (described below). A coherence value near 1.0 indicates a good test. In the regions where the value of coherence is low, something invalidated the test and the transfer function should be ignored at these frequencies.

Note that by using transfer function, the system's natural frequencies and damping factors can be accurately determined. The coherence at these frequencies indicates the level of confidence with which conclusions can be drawn.

Coherence

Coherence is, basically, a cause/effect parameter. It can only be implemented in a multichannel spectrum analyzer because it is necessary to simultaneously gather both the "cause" signal and the "effect" signal. Thus, if the coherence between two channels has a value of 1.0, all the signal of one channel was caused by the signal of the other channel in a linear fashion.

After obtaining a transfer function between two points on a machine component as described above, the correctness of the measured results must be decided. Several sources of error exist:

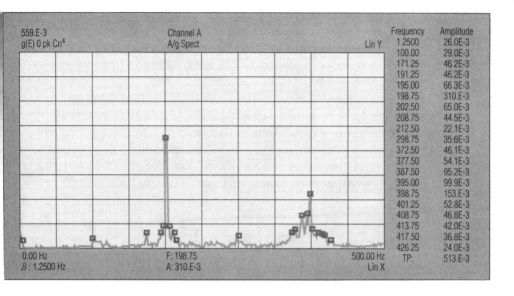

Figure F.3. Actual measured vibration of a dryer at present machine speed.

- Nonlinearity of the system under test
- Relatively high noise input levels that affect the output signal of the system but which were not part of the measured input signal
- Other system inputs that were not part of the measured input signal but which did affect the output signal (An example of this would be the vibration caused by people working on other parts of the paper machine during a shutdown)
- For impact tests, the failure to wait until the results of the first impact settle out before applying a successive impact (In this case, the residual of the first impact acts as an incoherent input during the second blow)
- Insufficient averaging
- System time delays that are large compared with the time window (1/bandwidth) chosen for the test

- Attenuation settings that do not take advantage of the full dynamic range of the analyzer, causing the internal noise of the analyzer to interference with one or both of the measured signals at frequencies where low energy exists.

It is extremely risky to draw conclusions from a transfer function test without checking the coherence display for an indication of one of the above errors. A low value of coherence at a frequency of interest should prompt a re-evaluation of testing procedure. Many spectrum analyzers have the ability to "blank out" those segments of a transfer function display having a coherence below a certain preset level.

Devising a Test

To avoid the great cost and lost production that can accompany the failed speedup of the twin-wire tissue machine that served as the impetus for the initial work, a plan was devised to determine whether the planned speed-ups of two fourdrinier machines was likely to cause excessive vibration problems. Such problems could occur in either of two ways:

1. The existing forcing mechanisms, such as misalignment, gear mesh frequencies, etc., will remain at the same or slightly higher amplitudes, but, since they will be rotating faster, may align with a structural resonance. If this happens, the vibration at that frequency will increase dramatically, possibly forcing a machine shutdown. This is the kind of problem the subject procedure attempts to anticipate.

2. Some forcing mechanisms, such as unbalance, can increase in amplitude due to a speed-up. Except for the specific case of unbalance, this kind of increase of amplitude is not highly predictable. The best that can be done in anticipation is to verify that the levels are of tolerable amplitude now and deal with a future problem in terms of improved balance or alignment criteria, or a component design change.

The new test procedure, then, has the goal of anticipating a change in the frequency of the forcing mechanisms known to exist in each piece of important rotating equipment in the paper machine and determining how that change affects the various measured transfer functions of the machine's structure. The problem of increases in the unbalance forces of the rotating members can be calculated in advance of the speed-up (if the transfer function is fairly flat in the region of running speed). Plans can be made to balance the offending rolls to tighter specifications well in advance of the scheduled speed-up date.

The Test Procedure

Fortunately, much of the testing that must be done can be performed without having an impact on current production rates. The procedure involves the following three phases:

1. The gathering of forcing frequency data
2. The determination of transfer functions
3. Data manipulation.

The following is a description of each of these phases. The required equipment necessary to proceed with each phase, and an estimate of the time and effort needed for each phase, are also given.

Gathering forcing frequency data

First, a test of each component of concern in the paper machine is carried out. This has the advantage of supplying actual frequencies and amplitudes under load. Work can be done with a magnetically mounted accelerometer and any 400-line FFT spectrum averager capable of performing at least eight summation averages on the frequency range of 0 to 500 Hz. Note that since low frequency data is needed, a velocity pickup is not an acceptable substitute for an ICP accelerometer with the capability of flat response down to 1 or 2 Hz.

The effort of gathering the data and, later, the data manipulation, will be greatly mitigated if an analyzer with the capability to store data on a PC format disk is used. Because a resonance can multiply a low amplitude signal by a factor of 100 or more and because of the high probability of an out-of-band signal reducing the effective dynamic range of the analyzer, it would be advisable to select an analyzer with a dynamic range of 80 db or more.

Not every component of the paper machine must be tested. If most of the felt rolls or a hitch roll were to cause a problem during a speed-up, for example, the solution would likely be simple and inexpensive. If, on the other hand, a dryer began vibrating excessively, a major problem would have befallen the project. In the case under discussion, data were gathered on each backside dryer bearing in the horizontal and vertical directions, as well as the backside bearings of some representative felt rolls. There is no need to measure vibration levels on those parts of the paper machine slated for replacement as a part of the speed-up program. It is important to note the speed of the machine while these readings are taken.

Because the afterdryers and afterdryer felt rolls had not been balanced in many years, vibration data were taken in the horizontal direction at each tending side dryer bearing and felt roll bearing to determine which rolls would have to be balanced as a result of the speed-up. The frequency range used was 0 to 50 Hz. These readings were necessary because the gear mesh signals on the backside of the machine were of high enough amplitude to mask the unbalance signal of most of the dryers because of dynamic range problems.

It should be noted that, since the forcing frequency data can be gathered at any known operating speed, the gathering of these data in no way hampers production. These measurements typically take a few days.

Transfer function data

The gathering of transfer function data is more difficult to accomplish. It must be done during a scheduled shutdown, with workers climbing all over the machine making various repairs and changing felts or wires. For this reason, the job must be carefully planned. Only typical rolls and bearing support/frames should be tested. The work should be laid out so that when most of the normal shutdown effort is directed at the wet end, dry-end transfer functions are obtained. Wet-end transfer functions must wait for dry-end repairs.

Obviously, the gathering of transfer functions requires a dual-channel spectrum analyzer. The job cannot be done with a single channel box. Aside from meeting the dynamic range capabilities mentioned above, the preview mode of averaging greatly speeds up the process. Also, the data manipulation is greatly simplified by having the transfer functions stored on a PC disk.

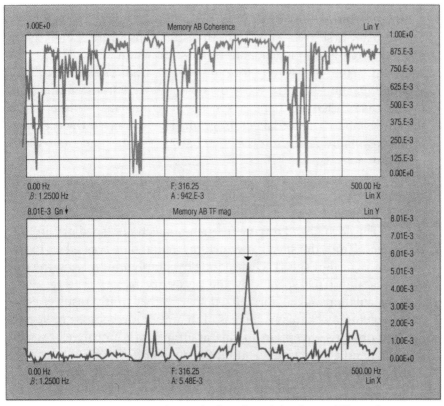

Figure F.4. Measured transfer function of the dryer of Figure 3.

It was found that an impact hammer of significant size must be used to "drown out" the work being done elsewhere on the machine. A 12-lb PCB force hammer was found to yield good coherence in the 0 to 500 Hz range for most test points. A 1-lb hammer was often inadequate. Since most rolls tend to be of similar design at both the tending and back sides of the machine, most of the transfer function work can be carried out on one side of the machine, reducing the number of instrument setups required.

Each transfer function must be accompanied by evidence that coherence of the test is acceptable. If two or three successive tests of a particular component show low coherence over a substantial region of the frequency range, it is best to forget it and go to the next component. Time is of the essence here because when the scheduled repairs have been made, the machine will be started up whether the vibration analyst is ready or not. With luck, sufficient transfer function data can be gathered during one or two 10- to 12-hour shutdowns.

Data manipulation

The first two speed-up investigations were done using plotted forcing frequency and transfer function data. Figure F.3 shows an actual measured vibration spectra from one of the rolls of a paper machine. Figure F.4 shows the relevant transfer function. In the case described here, a special program was written to do the entire job in a spreadsheet program. This increased the accuracy of the results by removing the "eye ball" estimates and the tedium from the analysis.

Each forcing frequency spectra is divided, frequency-by-frequency, by the relevant transfer function. This yields a normalized forcing frequency spectra.

The normalized forcing frequency spectra is "sped up" in 10% increments, from the speed at which the actual data was taken to the final future design speed. This is done by simply multiplying the frequency scale of the spectra by the necessary factors. The result of these calculations is the normalized spectra that would result if none of the forcing mechanisms increased in amplitude because of speed-up (the obvious exception to this assumption is unbalance, which must be handled separately) and if the transfer functions of the paper machine were flat.

Each of the sped-up, normalized spectra is multiplied by its corresponding transfer function (matching the new forcing spectra frequencies to the measured transfer function frequencies). This yields the expected sped-up forcing spectrum. The reason for speeding up the spectrum in 10% increments is to ensure that no major component goes through a resonance while coming up to the desired machine speed.

Figure F.5 shows the calculated spectra expected on the component of Figure F.3 after the speed-up. Having this spectra, it is an easy matter to decide if the levels are acceptable or not. Thus, each newly calculated spectra is compared with predetermined acceptability criteria for the particular component. Each exceedance is noted.

The only design changes required because of potentially high vibration levels are those for components that exceed the acceptability criteria. It is reasonably safe to assume that nonexceeding components will perform properly after speed-up. The few exceptions to this assumption are economically dealt with only after the problem has arisen, since the cause is likely to be nonlinear and incorrectly anticipated before speed-up.

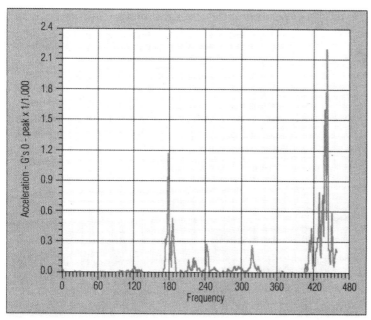

Figure F.5A Calculated vibration of the dryer of Figure F.3 to be expected during the first stage of speed-up.

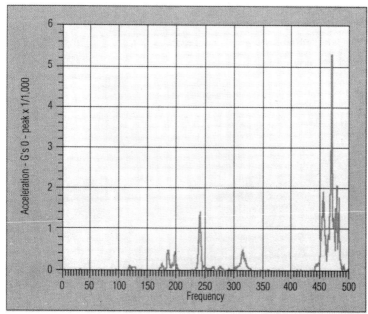

Figure F.5B Calculated vibration of the dryer of Figure F.3 to be expected during the second stage of speed-up.

Appendix F

Balance

As was pointed out above, unbalance is the most likely exception to the assumption that the sped-up forcing mechanisms will have the same amplitude as the measured spectra for a flat transfer function. Fortunately, the problem is easily dealt with.

First, measure the horizontal vibration at running speed for each of the rolls of interest. Then, increase the amplitude of that vibration by the square of the ratio of the anticipated speed to the measured speed. If the resultant level of vibration exceeds that deemed acceptable, the roll must be balanced to tighter tolerances before the speed-up takes place.

Summary

Using the above method of calculating anticipated vibration levels resulting after the paper machine speed-up by extrapolating current data allows those planning for a paper machine speed-up to anticipate future vibration problems. At the same time, as changes are made to the dryer section of the machine for increased capacity, and some larger drives are installed at various locations in the machine, structural changes or improved balancing can be made to the expected vibration trouble spots. Finally, correct the amplitude by the transfer function in increasing 10% speed increments. In this way, a far less costly and troublesome speed-up modification program can be expected.

APPENDIX G

Motor and Generator Vibration

Written By
Robert L. Wall
Apparatus Services, Inc.
Box 490 McGuire School Rd.
Delanson, NY 12053
518-875-8167

Introduction

The vibration spectra produced by rotating electric machines are similar in many respects to the vibration spectra produced by other rotating machines. Vibration frequencies caused by mechanical forces at running speed, and multiples of running speed vibration, as may be produced by rotor unbalance or the forces produced by shaft misalignment are the same. Vibration produced by bearings, such as the rolling element frequencies produced by rolling element bearings are also the same. But, in addition, rotating electrical machines produce a multitude of other vibration frequencies that are generated by the electromagnetic forces inherent in the operation of the machine.

Each machine, whether it is a motor or a generator, has its own individual vibration signature. The characteristic vibration signature is determined by the running speed, the number of poles on the rotor, the number of slots in the rotor and stator, and other variables. Certain vibration frequencies are fixed by the basic design and construction of the machine, but others may develop due to wear or deterioration of some components. This makes the monitoring of the vibration signature particularly useful for predictive or preventive maintenance purposes.

AC Motors and Generators

The stators of most modern motors and generators contain the armature winding, and the magnetic stator core provides part of the magnetic circuit of the machine. The stator core is made of thin steel laminations to control losses, and the bore of the stator has a number of uniformly distributed radial slots that hold the stator windings. The number of slots is determined by the size of the machine, number of poles, voltage, and other design variables.

The configuration of the rotor depends upon the type of machine: induction, salient pole synchronous or round rotor synchronous. In any case the rotor serves as the other part of the magnetic circuit and is separated from the stator by the air gap, a small radial clearance. The magnetic flux crossing this air gap produces forces that act to deform the rotor and stator and cause them to vibrate at a variety of frequencies and mode shapes.

Twice Line Frequency Vibration

The magnetic flux produced by current carrying conductors in AC machines alternates at line frequency. The force produced by the flux varies as the flux density squared. Therefore, if the flux is

represented as $B = B_o \sin wt$, then force $F = (B)^2 = (B_o \sin wt)^2$ and $F = B^2/2 (1 + \cos 2wt)$. The force has a steady component (a DC component as represented by the 1), and the twice line frequency component as represented by the cos2wt term. Therefore, all AC machines produce a twice line frequency vibration. In power systems that operate at 60 Hz the vibration frequency is 120 Hz (100 Hz in 50 HZ systems). In an induction motor, the poles in the stator winding produce a rotating magnetic force that acts across the air gap between the stator and rotor. This produces 120 Hz vibration of the stator, and the mode shape of the stator core vibration is determined by the number of poles in the winding. The number of poles in the winding and the line frequency also determines the synchronous speed of the rotating magnetic field and the running speed of the motor or generator.

RPM = 7200/number of poles (on a 60 Hz system)

Therefore a two pole winding produces a 3600 RPM motor, a 14 pole winding produces a 514 RPM motor.

A two pole winding has one north and one south pole, and produces a magnetic field as shown in Figure G.2.

The concentration of flux results in a force that acts to deform the stator into an elliptical pattern as shown in Figure G.3.

The two pole force wave deforms the stator into a mode shape having 4 nodes, a node being a point of zero deflection or vibration. Since the magnetic field is rotating at 3600 RPM the stator is deformed into a rotating elliptical pattern. It can be then visualized that a point of support for the stator core would be subjected to two vibration cycles for each revolution of the rotating ellipse.

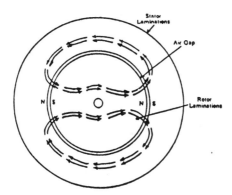

Figure G.2. Cross section of two pole induction motor showing magnetic circuit.

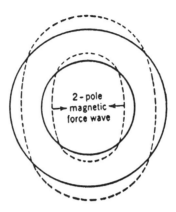

Figure G.3. Four node vibration pattern of stator core.

Similarly, a six pole motor produces a rotating magnetic field at 1200 RPM. The six pole force wave deforms the stator core into a mode shape having 12 nodes.

If the diameters of the two and six pole stators are the same, the circumferential span (pole pitch) of each pole will be different. The pole pitch of the two pole winding will be the circumference divided by two, while the pole pitch of the six pole winding will be the circumference divided by six. Similarly, the distance between nodes in the vibration pattern produced will be different. The distance between nodes will be the circumference divided by four and twelve for the two pole and six pole stators respectively. This is a very important consideration, because the deflection (vibration) of the stator core is analogous to the deflection of a beam. For a given load (force) the deflection decreases as the distance between beam supports is decreased.

The stator core is mounted in a frame that holds the core together and resists the reactive torque. It also provides the structure to maintain the air gap alignment and machine mechanical integrity. The points of support of the stator core transmit the radial 120 Hz vibration of the core to any attached structure such as bearing brackets, enclosures, and foundations. Harmonics of the 120 Hz can sometimes be encountered. Loose components or impacts produced by vibrating components can produce these harmonics.

As in any mechanical system, the amplitude of vibration produced is determined by the magnitude of the vibratory force and the dynamic response characteristics of the system. Separation of forcing frequencies and resonant frequencies is fundamental to realizing low vibration or noise levels. Stator cores are essentially thick rings and, as such, have a multitude of resonant frequencies. The first order resonant frequency of a ring has four nodes, the elliptical shape previously mentioned. As the order increases so do the number of nodes: 6, 8, 10, ... 24 . . . 56, etc. For a given size ring, the resonant frequency increases as the number of nodes increases. The resonant frequencies of the ring are functions of the mode shape, the diameter of the neutral axis, the sectional moment of inertia the cross section, the modulus of elasticity of the material, etc. For a given mode shape, the larger the diameter of the stator, the lower the resonant frequency. Also, for a given size machine, the higher the number of nodes in the force wave, the lower the vibratory response will be to the forces wave. High speed machines are much more likely to produce high levels of 120 Hz vibration than low speed machines.

The above discussion assumes that the air gap is uniform. In the real world the air gap is never perfectly uniform, and to a matter of degree the air gap is asymmetrical.

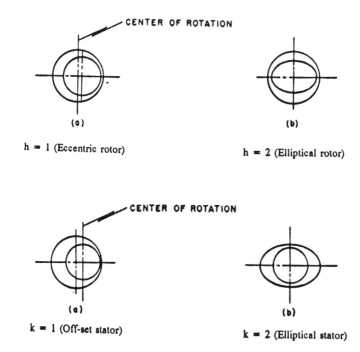

Figure G.4. *Low order air gap dissymmetries.*

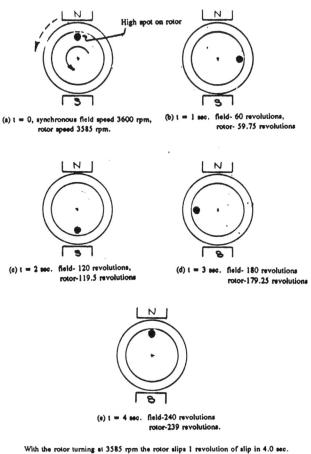

With the rotor turning at 3585 rpm the rotor slips 1 revolution of slip in 4.0 sec. with respect to the rotoating synchronous field. The high spot on the rotor comes under the influence of a stronger magnetic pull at the center of a pole twicw per revolution of slip.

SLIP FREQUENCY = 15 CPM OR 0.25 CPS. MODULATION FREQUENCY OF 120 HZ. STATOR CORE VIBRATION = 30 CPM OR 0.50 CPS

Figure G.5. Mechanism of twice slip frequency modulation of twice line frequency stator core vibration.

Figure G.4 shows typical, low order air gap dissymmetries. The dissymmetries may be due to the accumulation of tolerances during manufacture or they may be due to wear or poor maintenance practices. The most common air gap dissymmetry is the rotor offset with respect to the stator bore. In normal operation there is a considerable force that is acting to collapse the stator and expand the rotor. This force may be in the range of 35 pounds per square inch. The force is a function of the flux density of the air gap, but also the length of the air gap. The smaller the gap, the higher the force. If the rotor is offset in the bore of the stator, an unbalanced magnetic force is created which acts to pull the rotor and the stator together. In some situations this actually happens and the rotor rubs the stator and causes serious (if not catastrophic) damage.

The unbalanced magnetic force is a steady or DC force, but it also has a 120 Hz component. The magnitude of the 120 Hz component is a function of the magnitude of the unbalanced force and the number of poles in the winding. As one would intuitively expect, the higher the number of poles, the lower the 120 Hz component. The unbalanced force depends upon the size, the design characteristics of the machine, and the amount of air gap dissymmetry. In large machines, the DC component can be several thousand pounds.

The 120 Hz component produces a 2 node vibration pattern. It acts to vibrate the rotor and stator back and forth. This frequently can result in excessive vibration, especially on vertical machines.

If the air gap dissymmetry is caused by the rotor, a different behavior will be evident. For instance, if the rotor body of an induction motor is eccentric relative to its center of rotation, the point of minimum air gap will rotate at rotor speed which is less than the

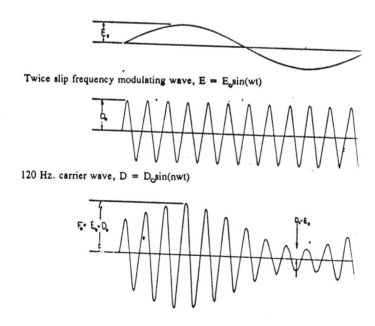

Twice slip frequency modulating wave, $E = E_o \sin(wt)$

120 Hz. carrier wave, $D = D_o \sin(nwt)$

Twice slip frequency modulated 120 Hz. Vibration $F = ((D_o + E_o\sin(wt))\sin(nwt)$

Figure G.6. *Amplitude modulated vibration in real time.*

synchronous speed of rotation of the magnetic field. (A broken rotor bar will produce a magnetic dissymmetry that produces the same affect). As the point of minimum air gap rotates, it slips past the magnetic poles. At the center of the poles the magnetic pull is maximum; it is minimum between the poles. This is illustrated schematically in Figure G.5.

This produces an amplitude modulation of the 120 Hz core vibration as shown in Figure 5.

This is the cause of the typical pulsating noise or vibration that is often associated with large 2 pole, 3600 RPM induction motors.

Current carrying conductors create their own magnetic fields and the 120 Hz forces that can cause vibration of nearby magnetic components. For instance air baffles in close proximity to stator windings may be caused to vibrate excessively. Rotor fans can experience similar problems during starting of motors.

Stator windings are current carrying conductors in magnetic fields and are subjected to relatively high levels of 120 Hz forces. Properly wedged and braced windings will not vibrate excessively, but loose windings can vibrate excessively and winding failure may result due to the abrasion of insulation or the fatigue of the conductors.

Changes in the magnitude of 120 Hz vibration may be an early warning sign that looseness has developed and corrective action is necessary. Loose components driven to vibrate by 120 Hz forces may cause impacts that produce vibration rich in harmonics of 120 Hz.

Slot Frequency Vibration of Induction Motors

The stator windings generate the magnetomotive force (MMF) that produces the flux that drives the motor. In addition to the fundamental flux at line frequency, the windings also produce harmonics of line frequency (space or phase belt harmonics and saturation harmonics). The presence of slots on the surfaces of the rotor and stator produce permeance variations in the air gap.

Figure G.7 shows typical rotor and stator slots. The air gap flux is a product of the MMF times the permeance. The forces acting across the air gap are a function of the flux density squared. The MMF and the permeance variations can be represented by a series of sine and cosine waves, as can be the force waves. Carrying out the mathematics is laborious and results in a multitude of force waves having a wide variety of frequencies and numbers of poles. The series can be shortened considerably by eliminating all force waves except those having a low number of poles (relatively long pole pitch). The stator core response to very short pole pitches is minimal and, therefore, no significant vibration is produced.

Motor and Generator Vibration 277

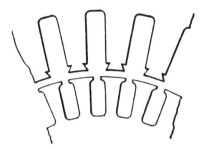

Figure G.7. Section of rotor and stator cores showing typical slots.

The table in Figure G.8 lists the principle force waves produced by an induction motor with a symmetrical air gap. Given the slot combination, the number of rotor and stator slots, and the number of poles in the stator winding, the principal force waves can be calculated using the equations shown. Equations are given for motors having a greater number of stator slots than rotor slots, and for the opposite condition. Note that the frequency of the force waves varies as "s", the amount of slip. The rotors of induction motors rotate at less than the synchronous speed of the rotating magnetic field. The difference between synchronous speed and actual running speed is the slip. For these purposes, the slip is expressed in per unit. At rotor standstill the slip is 1.0. For example, if the rotor of a 4 pole motor is operating at 1770 rpm on a 60 Hz power supply, the slip is 1800 rpm –1770 rpm or 30 rpm. The per unit slip is 30/1800 or .0167.

The affect of a nonuniform air gap on the slot frequency vibrations is similar to that produced on the 120 Hz vibration. Nonrotating dissynmmetries, i.e. the rotor offset relative to the stator or an elliptical stator bore, produce force waves having the same frequencies, but different numbers of poles. Dissymmetries of the rotor that cause rotating air gap dissymmetries produce new force waves having different frequencies as well as different numbers of poles. See Figure G.4 for typical, low order air gap dissymmetries. The tables shown in Figures G.9 and G.10 give the equations for calculating the force waves created by different air gap dissymmetries. Note that these force waves are in addition to those calculated for motors with uniform air gaps.

Figure 10 shows a typical vibration frequency spectrum for a large, 6 pole induction motor. In this case the stator had 90 slots, and the rotor had 76 slots. The primary force waves as calculated using the formulas in Figure G.8 occur at 1400, 1520, and 1640 Hz at no load

FORMULAS FOR CALCULATING VIBRATION PRODUCING MAGNETIC FORCE WAVES IN INDUCTION MOTORS

(FOR 3 PHASE MOTORS HAVING 60 PHASE BELTS)

SOURCE HARMONIC	S > R, STATOR SLOTS MORE NUMEROUS THAN ROTOR SLOTS		R > S, ROTOR SLOTS MORE NUMEROUS THAN STATOR SLOTS	
	NUMBER OF PAIRS OF POLES	FREQUENCY FACTOR	NUMBER OF PAIRS OF POLES	FREQUENCY FACTOR
1M	S - R + 2P	r - 2	R - S + 2P	r + 2
1M	S - R	r	R - S	r
1M	S - R - 2P	r + 2	R - S - 2P	r - 2
5M	S - R - 4P	r - 2	R - S - 4P	r + 2
5M	S - R - 6P	r	R - S - 6P	r
7M	S - R - 6P	r	R - S - 6P	r
7M	S - R - 8P	r + 2	R - S - 8P	r + 2
3N	S - R - 2P	r + 2	R - S - 2P	r - 2
3N	S - R - 4P	r + 4	R - S - 4P	r - 4
5N	S - R - 4P	r + 4	R - S - 4P	r - 4
5N	S - R - 6P	r + 6	R - S - 6P	r + 6

NOMENCLATURE: $P = 1/2$ the number of poles in the stator winding, R = number of rotor slots, S = number of stator slots, $r = R/P(1-s)$ where s = per unit slip, M = space harmonic, and N = saturation harmonic. Note: Actual magnitude of M and N harmonics depend upon motor design and operating conditions.

To determine actual frequency of force wave multiply frequency factor by line frequency in Hz.

Figure G.8. Formulas for calculating number of poles and frequencies of principle force waves in induction motors.

(essentially 0 slip). When the motor is loaded, the rotor rpm will decrease slightly and cause a small decrease in these frequencies.

Additional force waves due to dissymmetries calculated using Figure G.9 occur at several other frequencies in the 3000 Hz range with side bands of running speed present.

As can be seen, many frequencies occur due to the multiplicity of force waves present. The vibration response to these forces depends upon the resonant frequencies of the motor components, including localized resonances of panels, etc.

FORMULAS FOR CALCULATING VIBRATION PRODUCING FORCE WAVES IN INDUCTION MOTORS DUE TO ROTOR AND STATOR DISSYMMETRIES

(FOR 3 PHASE MOTORS HAVING 60 PHASE BELTS)

SOURCE HARMONIC	S > R, STATOR SLOTS MORE NUMEROUS THAN ROTOR SLOTS	
	NUMBER OF PAIRS OF POLES	FREQUENCY FACTOR
1M	(mS - jR + 2P +/- k +/-h)	(jR - 2)f +/- hQ
1M	(mS - jR +/- k +/-h)	(jR)f +/- hQ
1M	(ms - jR - 2P +/- k +/-h)	(jR + 2)f +/- hQ
5M	(mS - jR - 4P +/- k +/-h)	(jR - 2)f +/- hQ
5M	(mS - jR - 6P +/- k +/-h)	(jR)f +/- hQ
7M	(mS - jR - 6P +/- k +/-h)	(jR)f +/- hQ
7M	(mS - jR - 8P +/- k +/-h)	(jR + 2)f +/- hQ
3N	(mS - jR - 2P +/- k +/-h)	(jR + 2)f +/- hQ
3N	(mS - jR - 4P +/- k +/-h)	(jR + 4)f +/- hQ
5N	(mS - jR - 4P +/- k +/-h)	(jR + 4)f +/- hQ
5N	(mS - jR - 6P +/- k +/-h)	(jR + 6)f +/- hQ
7N	(mS - jR - 6P +/- k +/-h)	(jR + 6)f +/- hQ
7N	(mS - jR - 8P +/- k +/-h)	(jR + 8)f +/- hQ

NOMENCLATURE: P = 1/2 the number of poles in the stator winding, R = the number of rotor slots, S = the number of stator slots, r = R/P(1-s) where s = per unit slip, M = space harmonic, N = saturation harmonic, (Note; actual magnitude of M and N harmonics depend upon motor design and operating conditions), m and j are orders of stator and rotor slot symmetries, ie 1 if slots and teeth are of equal width, and 2 if widths are not equal; h and k are rotor and stator dissymmetries as shown in Figure 3.

To determine actual freqeuncies in Hz. multiply frequency factors by line frequency in Hz. and add or subtract multiples of running speed in RPS depending upon the order of the rotor dissymmetry.

Figure G.9. *Formulas for calculating the number of poles and frequencies of principle force waves in induction motors due to rotor and stator dissymmetries.*

Stator Core Resonant Frequencies

The level of vibration or noise that any one of the multitude of electromagnetic force waves will produce depends upon the magnitude of the force wave, but also upon the dynamic response characteristics of the stator core. The magnitude of many of the force waves

FORMULAS FOR CALCULATING VIBRATION PRODUCING FORCE WAVES IN INDUCTION MOTORS DUE TO ROTOR AND STATOR DISSYMMETRIES

(FOR 3 PHASE MOTORS HAVING 60 PHASE BELTS)

SOURCE HARMONIC	R > S, STATOR SLOTS MORE NUMEROUS THAN ROTOR SLOTS	
	NUMBER OF PAIRS OF POLES	FREQUENCY FACTOR
1M	(jR - mS + 2P +/- k +/-h)	(jR + 2)f +/- hQ
1M	(jR - mS +/- k +/-h)	(jR)f +/- hQ
1M	(jR - mS - 2P +/- k +/-h)	(jR - 2)f +/- hQ
5M	(jR - mS - 4P +/- k +/-h)	(jR + 2)f +/- hQ
5M	(jR - mS - 6P +/- k +/-h)	(jR)f +/- hQ
7M	(jR - mS - 6P +/- k +/-h)	(jR)f +/- hQ
7M	(jR - mS - 8P +/- k +/-h)	(jR - 2)f +/- hQ
3N	(jR - mS - 2P +/- k +/-h)	(jR - 2)f +/- hQ
3N	(jR - mS - 4P +/- k +/-h)	(jR - 4)f +/- hQ
5N	(jR - mS - 4P +/- k +/-h)	(jR - 4)f +/- hQ
5N	(jR - mS - 6P +/- k +/-h)	(jR - 6)f +/- hQ
7N	(jR - mS - 6P +/- k +/-h)	(jR - 6)f +/- hQ
7N	(jR - mS - 8P +/- k +/-h)	(jR - 8)f +/- hQ

NOMENCLATURE: P = 1/2 the number of poles in the stator winding, R = the number of rotor slots, S = the number of stator slots, r = R/P(1-s) where s = per unit slip, M = space harmonic, N = saturation harmonic, (Note; actual magnitude of M and N harmonics depend upon motor design and operating conditions), m and j are orders of stator and rotor slot symmetries, ie 1 if slots and teeth are of equal width, and 2 if widths are not equal; h and k are rotor and stator dissymmetries as shown in Figure 3.

To determine actual freqeuncies in Hz. multiply frequency factors by line frequency in Hz. and add or subtract multiples of running speed in RPS depending upon the order of the rotor dissymmetry.

Figure G.10. Formulas for calculating the number of poles and frequencies of principle force waves in induction motors due to rotor and stator dissymmetries.

is quite small. Excessive vibration is often the result of the coincidence, or near coincidence, of the frequency of a force wave and a resonant frequency of the stator core. Resonant amplification is most significant when the number of poles in the force wave coincides with the mode shape of the resonant frequency, for example, when the force wave having 12 poles and its frequency coincides with a stator core resonance having 12 nodes at the same frequency.

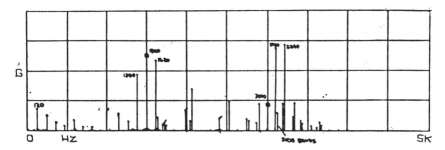

Figure G.11. *Vibration frequency spectrum of a large 6 pole induction motor with 90 stator slots and 76 rotor slots.*

The calculation of the resonant frequencies of a stator core is difficult and inexact in many cases because of the many variables involved. In addition, mechanical dissymmetries in the core or frame may produce several resonant frequencies with the same mode shape. It is beyond the scope of this text to discuss the subject in detail, but there are some factors that affect the resonant frequencies of the core that can change and significantly alter the core resonant frequencies.

As in any mechanical system, the resonant frequency is a function of the square root of the ratio of the stiffness to the mass of the system. The stator core consists of a yoke which is a thick walled cylinder with slots on the inside diameter. The slots hold the windings. The windings, and the teeth formed by the slots, add mass, but not necessarily proportional stiffness to the core. In addition, the stiffness of the frame in which the core is mounted, and the method of attachment of the core to the frame, add other difficult-to-evaluate variables. Because of the nature of these variables, they can change over a period of time due to the effects of temperature, wear, and abrasion, etc., and cause a change in the vibration of the motor.

The stiffness of the core is a function of the modulus of elasticity of the core material, among other things. The core is made up of thin laminations of silicon steel. Depending upon motor size, age, and manufacturer, the laminations may be one-piece complete circles or they may be segments of a circle assembled to form a complete circle. In the case of a segmented core, the effective modulus of elasticity varies with the axial pressure holding the laminations together. Consequently, as cores and their clamping arrangements relax with time and temperature, the effective modulus is reduced which, in turn, lowers the resonant frequencies. This may explain why the vibration or noise produced by a motor may change with everything else being equal.

An increase in the resonant frequencies of the core may occur if the core is tightened or vacuum pressure impregnated with a resin that bonds the laminations together, which would act to increase the stiffness.

Changes can also be produced if the stator winding is replaced with windings having properties different from the original windings. For instance, replacing an original winding, which had soft insulation without vacuum pressure impregnation treatment, with a winding which is vacuum pressure impregnated with a high-strength resin, will produce a stator having different characteristics. Better coupling between the windings and the core increase the effective mass of the core, but better fill of the slots may increase the effective stiffness.

Fretting corrosion between the core and the method of attachment of the core to the frame may occur due to long term relative vibration. Decoupling of the core from the frame may also significantly change the characteristics of the core/frame system.

The above are discussed in a qualitative sense only and presented to show that motor resonant frequencies can change. The amount of noise produced may increase or decrease as a result. Recognition of these causes may help to understand a change in signature which may lead to more in depth investigation and analysis of the machine condition.

Rotor Winding Effects on Slot Frequency Vibration and Noise

The effects of air gap dissymmetries on 120 Hz and slot frequency vibrations were previously discussed. However, the air gap may be mechanically symmetrical and similar effects can be caused by anomalies in the rotor windings. The squirrel cage windings of induction motors are subjected to significant thermal expansion and centrifugal forces. The cyclic stresses produced by starting and stopping will eventually result in metal fatigue failure of the bars or endrings of the winding. The broken bars create a dissymmetry in the air gap flux pattern that produce new force waves with different frequencies that can be detected with narrow band width filters. The frequencies and numbers of poles in these force waves can be calculated using the formulas given in Figures G.9 and G.10.

Slot Frequency Vibration of Salient Pole Synchronous Machines

Salient pole synchronous machines sometimes produce vibration and noise in the middle-to-high audible frequency range, not unlike that produced by induction motors. The stators of synchronous machines are for all practical purposes identical to those of induction motors. The stator has a certain number of slots as determined by the size, speed, and voltage of the machine. The stator is connected to produce a rotating electromagnetic field where the speed of rotation is determined by the number of poles in the winding and the frequency of the power supply.

The rotor has a number of salient poles equal to the number of poles in the stator winding. The poles are connected to a source of DC power in a way to produce alternate north and south poles. The magnetic flux crossing the air gap produces a series of force waves that have a multitude of frequencies and numbers of poles similar to those in the induction motor. The analysis to define the force waves is similar to that described for induction motors. Figure G.12 shows a schematic section of a synchronous machine.

Carrying out the mathematics is laborious, yielding a series of force waves having different frequencies and numbers of poles. Again, the series can be simplified by considering only those force waves having a relatively few number of poles (relatively long pole pitch).

The following equation can be used to calculate the frequencies and number of poles in the air gap force waves.

$$F = \cos[(np-mS) \times -npNt]$$

where n is an integer odd or even, and m is any integer, but in most practical machines m=1 or 2. If the width of the slots and the teeth are about equal, m=1. If the slot is much wider than the teeth or vice versa, m=2. S = the number of stator slots, p = the number of pairs of poles in the machine (i.e., for a 12 pole machine p=6), and N is the rotational speed n radians per second.

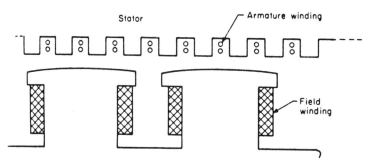

Figure G.12. Section of synchronous motor showing salient field poles and stator slots.

284 *Appendix G*

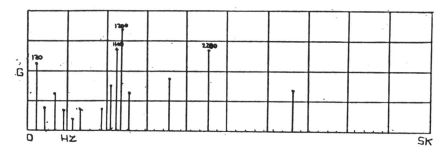

Figure G.13. Vibration frequency spectrum of a 10 pole synchronous motor with 90 stator slots.

np-mS is the number of pairs of poles in the force wave, and npN/6.28 is the frequency of the force wave in Hz.

The most important force waves (having the long pole pitches) will generally occur when n is equal to or close to the number of slots per pole in the stator winding, or multiples of this value.

Figure G.13 shows a vibration spectrum of a 10 pole synchronous motor that has 90 stator slots, or 9 slots per pole. For a value of n = 9, a force wave with 45 pairs of poles at 540 Hz is produced. Force waves with so many poles generally will produce little response. The following table shows the force waves for various values of n, and m = 1.

n	Pairs of poles in force waves	Frequency
9	45	540 Hz
10	40	600
16	10	960
17	5	1020
18	0	1080
19	5	1140
20	10	1200
With m = 2	mS = 180	
36	0	2160
38	10	2280

It is important to identify the source of the various frequencies produced by the motor so that they can be distinguished from frequencies produced by other sources.

Influence of Air Gap Dissymmetries

If the air gap between the rotor and stator is not symmetrical, additional force waves will be produced in a manner similar to that for induction motors (see Figure G.4). Force waves having different numbers of poles and different frequencies will be produced depending on the type and order of the dissymmetry.

The number of poles in the force waves produced by the air gap dissymmetry in synchronous motor is:

(np–mS) +/–k where k is the order of the dissymmetry, k= 1 for offset air gap and k = 2 for elliptical air gap.

The frequencies of the force waves remain the same as without the dissymmetries: npN/6.28.

Influence of MMF Dissymmetries

If dissymmetries associated with the MMFs produced by the individual field poles are present, then a rotating dissymmetry is produced. This may be caused by design or manufacturing irregularities or, more likely, by the development of shorted turns in the windings of one or more of the field poles. The shorted turns reduce the flux produced and a nonuniform flux pattern results. The rotating dissymmetry can be represented by additional terms in the basic equations, but they can be reduced to those terms producing force waves with a low number of poles. Because the dissymmetry is rotating, it produces additional force waves having different frequencies, and numbers of poles.

The equations for these force waves are:

$$(n-mS) +/-j = \text{the number of poles}$$

$$(n +/-j) \, N/6.28 = \text{frequencies in Hz}$$

Shorted turns can also produce an unbalanced magnetic pull that acts on the rotor in a way similar to mechanical unbalance. Depending upon the phase relationship with the mechanical unbalance, it can either increase or decrease the rotor vibration at one times running speed. The unbalanced magnetic force can be canceled out by the addition of balance weights of the proper weights and locations.

Summary

Rotating electric machines produce a multitude of vibration frequencies that are inherent in the design and operation of the machine. As has been shown, the magnitude of vibration at any given frequency depends upon the number of poles in the force wave acting across the air gap, and the dynamic response characteristics of the major and minor machine components. Changes can occur within a machine that may change the nature of the force waves produced and(or) the dynamic response characteristics of the machine components. It is therefore desirable to monitor the vibration spectrum of key machines in order to identify possible changes that may be indicative of developing problems caused by wear, looseness, or other causes of mechanical deterioration.

APPENDIX H

Data Collector, Spectrum Analyzer Operational Verification Procedures

Written by
John Visotsky
V-TEK Associates
Marietta, GA

There are many stories of catastrophic machine failures that could have been avoided if a vibration monitoring program was in place. Sometimes notes show that vibration data was collected. The vibration analyst and software trending program found no problems. Could the problem have been a defective analyzer that was not operating properly or did not meet its published specifications? If the analyzer is not properly acquiring vibration data, the trending software and analyst can not make an accurate assessment of machine condition.

"Is my data collector/analyzer functioning properly? Is it operating within its stated performance specifications? Does the internal calibration signal perform an adequate system conformance test of my analyzer?" Anyone who uses a data collector or FFT analyzer to monitor the condition of critical machinery should ask these questions and periodically verify the integrity of the analyzer used to perform these tests.

The implementation of an analyzer operational verification procedure or a complete calibration and test to the manufacturer's published specifications should be carried out on a regular basis. If the analyzer is sent to a calibration lab, the calibrations should be traceable to the National Institute for Standards and Technology (NIST), and performed in accordance with MIL-STD-45662A.

This appendix will guide us through a number of key tests that should be the core of an operational verification procedure.

Most of today's data collectors, specifically FFT spectrum analyzers, are designed to provide the user with a multitude of analysis settings. The most commonly used spectrum analyzer settings are input sensitivity, sensor calibration, frequency analysis range, digital filter selection, and display mode.

We can appreciate the complexity of an FFT spectrum analyzer by reviewing these most commonly used settings. Input sensitivity can typically be set in 10 or more steps from typical ranges of 10 millivolts to 10 volts. These sensitivity ranges can be used in combination with frequency analysis ranges starting at DC to 10 Hertz (600 cpm) selectable in 10 or more steps up to DC to 100,000 Hertz (6,000,000 cpm). As you can see, by just selecting input coupling, sensitivity and a frequency analysis range, the operator can enable any one of more than 100 combinations of analysis settings.

Each of these settings access a unique analog circuit and digital hardware signal path required to route an input signal for additional processing within the analyzer prior to presenting the completed process function on the analyzer's display screen.

The block diagram below is representative of a typical data collector/analyzer. This diagram may include certain functions not found on all analyzers and omit or alter the implementation of functions present on some analyzers. The purpose of the diagram is to give the reader an idea of the signal flow from the input of the analyzer to the final display presentation.

Required Test Equipment

Operational verification will require external test equipment. The recommended test equipment listed below is based on the assumption that the data collector or FFT analyzer under test does not have sample rates that exceed 256 KHz.

- Frequency counter, DC to 300 KHz (.001% accuracy)*
- Digital multi-meter (0.01% AC accuracy 3 to 100 Hz)*
- Low distortion (<96 dBc) function generator, DC to 300 KHz
- Cables, attenuators and terminators
- Portable shaker for vibration sensor calibration

---

*Typically, frequency and amplitude accuracy of the calibration test equipment should be ten times greater than the specifications of the unit under test.

Operational Verification Procedures 289

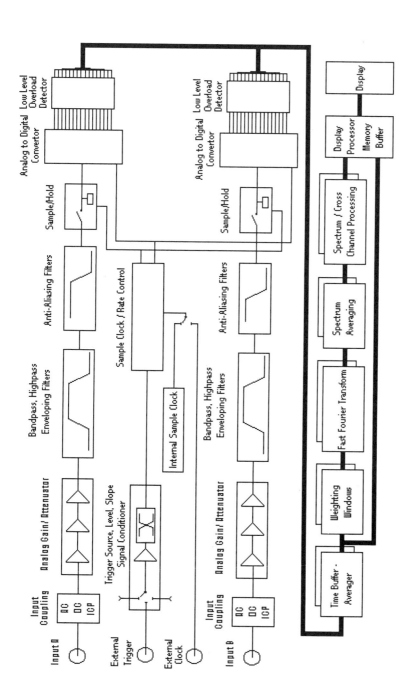

Figure H.1. A typical data collector/analyzer.

The tests listed below cover key specifications common to most FFT spectrum analyzers and should be the core of an operational verification procedure.

- Amplitude accuracy
- Frequency accuracy
- Antialiasing filter passband flatness
- ICP voltage, current test and accelerometer calibration
- Dynamic range, amplitude linearity, spurious and harmonic components
- Antialiasing filter stopband attenuation.
- Transfer function, gain and phase accuracy

Initial Setup and Amplitude Accuracy

Prior to the start of the operational verification procedure, the analyzer under test must be preset to the following initial settings.

Input coupling set to AC, vertical amplitude scaling set to RMS Volts/Log display, Integration off, time and spectrum averaging off, trigger to freerun and all other special filters and functions off. Set input sensitivity to 2.0 Volts, frequency display range to 1.0 KHz and resolution to 400 line spectra. Set Windowing to Flattop weighting for maximun amplitude accuracy. The objective of this test is to verify amplitude accuracy of all gain and attenuator input combinations. Note: Typically the total sensitivity range is about 130 dB, limited to the highest amplitude range and the system noise floor. Verification of these specifications can be included at this point in the test procedure.

The next step is to set the external signal source output to 0.700 Volts RMS at 100 Hz. Connect the output to the analyzer under test and an external RMS voltmeter. Be sure to use a proper terminator, typically a 50 ohm resistor. Adjust the external frequency source output level to read exactly 0.700 Volts RMS on the external RMS voltmeter. Then note the amplitude measured at 100 Hz on the analyzer under test. The analyzer amplitude display at 100 Hz should read between 0.693 and 0.707 Volts RMS if the accuracy specification is +/- 1.0%. A sample data sheet is shown below.

Amplitude Accuracy +/- 1.0% (abbreviated sample, gain/attenuator section)

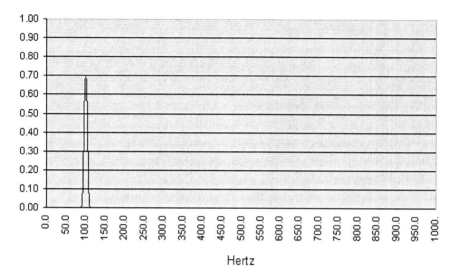

Figure H.2. Sample graph of the amplitude accuracy test with the analyzer sensitivity set to 1.00 Volts and an input signal of 0.700 Volts RMS.

Sensitivity Range	Input Level	Specification	Measured Level	Status
10.0 Volts	7.000 Vrms	6.930>7.070	7.002	Pass
5.0 Volts	3.500 Vrms	3.465>3.535	3.490	Pass
2.0 Volts	1.400 Vrms	1.386>1.414	1.400	Pass
1.0 Volts	0.700 Vrms	0.693>0.707	0.701	Pass

Frequency Accuracy

The frequency accuracy of a data collector/analyzer can be checked by measuring the frequency of the sample clock with an external frequency counter. In many cases the sample clock is not accessible. An alternative method is to input a calibrated frequency and set the unit under test to the highest frequency resolution. If frequency accuracy is listed as +/- 0.1%, the following procedure may be used to verify compliance with this specification.

If a 100.00 Hz signal is input to the analyzer, the unit should measure between 99.90 Hz and 100.10 Hz. Set the analyzer to 200 Hz analysis range, Hanning weighting, and the maximum number of FFT lines to obtain the highest frequency accuracy and resolution. A 3200 line FFT will yield 0.0625 Hz resolution (200 Hz divided by 3200 lines).

Antialiasing Filter Passband Flatness

The Accuracy measurements above have tested only one input gain setting and one frequency line of a 400/800/1600 line, or greater, frequency span. Also there may be ten or more frequency spans. The objective of this test should be to get a reasonable indication that most combinations of the analog gain section and antialiasing filters are flat and within specification across all frequency spans. The second part of the amplitude accuracy test can be set up to monitor the flatness of the antialiasing filter. This can be accomplished by entering 10 or more individual frequencies for each analysis range from 1% to 99% of the Fmax for each analysis range. The objective is to get a reasonable profile of filter flatness without testing everyone of the 400 or more lines. A test data sheet and sample graphs for these tests can be configured as follows.

Amplitude Accuracy +/- 1.0% (abbreviated sample, antialiasing filter section) sensitivity range set to 1.0 Volts, input level set .700 Vrms at 10 points in the frequency spectrum, from .01 to .99 of Fmax. specification is 0.693>0.707

A/R(Fmax)	.01	.15	.25	.35	.45	.55	.65	.75	.85	.99	Status
50 KHz	.701	.701	.703	.705	.707	.705	.702	.699	.698	.698	Pass
20 KHz	.700	.699	.700	.701	.702	.701	.700	.699	.699	.700	Pass
10 KHz	.701	.703	.705	.707	.705	.702	.699	.698	.698	.699	Pass
5 KHz	.699	.700	.703	.705	.707	.706	.707	.706	.704	.701	Pass

Note: System amplitude accuracy is based on the accuracy of the analyzer combined with the sensor and cabling used in the measurement. Also, the lower frequency accuracy of an analog RMS voltmeter may be limited to 20 Hz, and, if AC coupled, the analyzer low frequency is limited to about 0.5 to 1.0 Hz. DC coupling should be used to measure oscillator frequencies below 1 Hz.

ICP Voltage, Current Test and Accelerometer Calibration

An accelerometer should also be tested to insure a complete operational verification of the system originating at the signal pickup sensor. Set the analyzer to ICP coupling and vertical calibration to the stated accelerometer sensitivity, typically 10 to 500 mVolt/g. Prior to connecting the sensor, measure the DC voltage and current present at

Operational Verification Procedures 293

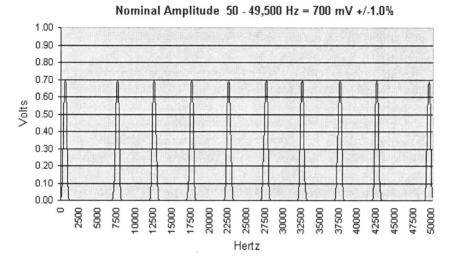

Figure H.3. Sample graph of the amplitude accuracy test with the analyzer sensitivity set to 1.0 Volts and multiple input signals of 0.700 Volts RMS stored at ten discrete frequencies across the frequency span using the peak hold averager mode.

the input BNC. The DC voltage should typically be 18 to 30 Volts and the constant current source is typically 2 to 20 mA. Enter the accelerometer's calibration factor and units of measure, (RMS or Peak), mount it on the shaker and connect it to the analyzer. Typical output is 1.0 g RMS at 159.2 Hz. The analyzer under test should read 1.0 g RMS at 159.2 Hz (+/- the accumulated tolerances of the shaker, sensor and analyzer.)

Figure H.4. PCB – IMI Model 394C04 Handheld Accelerometer Calibrator

Dynamic Range, Amplitude Linearity, Spurious and Harmonic Components

Dynamic range, linearity, spurious and harmonic component tests can be performed together since we will start with a near full scale amplitude signal noting the spectral purity of this signal, then, in steps, reduce the amplitude to the minimum detectable level. Dynamic range is not determined by the number of bits in the A/D nor how small a signal can be detected below full scale. Dynamic range is the difference between the amplitude of the highest signal input level (prior to the onset of an analyzer overload condition) and the lowest detectable signal of a selected sensitivity range.

The graph below displays a full scale signal at 100 Hz in a 1000 Hz frequency span. Selecting a low frequency input in this analysis range enables display of multiple harmonic and spurious components. In this sample test, the highest distortion component we see is a second harmonic that is 75 dB below the fundamental signal. Linear or RMS averaging can also be used to lower the noise floor and improve dynamic range.

In this test, the difference in amplitude between the signal set to maximum amplitude prior to overload at 100 HZ and the highest harmonic component, located at 200 Hz, is 75 dB. This test shows all other harmonic and spurious frequency components to be lower than 75 dB below the 100 Hz fundamental frequency for this gain setting

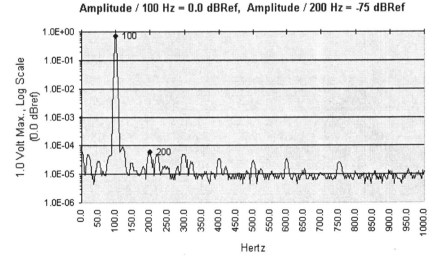

Figure H.5. A 100 Hz Signal

and analysis range. Therefore our dynamic range for this setting is limited to 75 dB due to the harmonics present in the display which are a product of the analyzer and not the signal source.

Antialiasing Filter Stopband Attenuation

This is a good point to test the stop-band attenuation of the antialiasing filter. The primary source of aliasing in analog to digital conversions is difference frequency between the input signal and sample clock. As an example, if the analysis range is set to 1 KHz with the sample rate at 2.56 KHz and an input frequency set to 2 KHz, the difference frequency is 2.56 KHz minus 2.00 KHz. This resulting difference frequency is 0.56 KHz which falls within the analysis range. If a signal appears within the specified dynamic range at 0.56 KHz, it is an alias frequency and an indication that the antialiasing filter is not functioning properly. All analyzers may not use sample clocks that are exactly 2.56 times the highest analysis range frequency. It may be necessary to try other out-of-band frequencies to test aliasing. An easy way to detect an alias is to look for an inband frequency that decreases as the out of band frequency is increased.

Amplitude Linearity Accuracy can be tested by changing the amplitude of the 100 Hz input signal in 10 dB increments and noting the difference between the measured input level and the level displayed on the analyzer. Amplitude accuracy should be maintained throughout the dynamic range.

Transfer Function, Gain and Phase Accuracy

A Transfer Function measurement should be performed on several analysis ranges of two channel analyzers to test phase and amplitude match between the channels. This test can be performed in a similar manner to the flatness test but, instead of measuring the amplitude at 10 points in the analysis range, we connect the same signal to both channels and measure Transfer Function Phase and Amplitude match between the two channels at the 10 measurement points. Typical specifications are +/- 1.0 degrees for phase and +/- 0.5 dB for amplitude match. This test should be performed for each analysis range. The Coherence function can also be verified in this test and should be equal to one for each of the 10 measurement points.

Summary

It can be concluded that a periodic test of the data collector or analyzer is of equal or greater importance than monitoring the condition of any single critical machine in the plant. Most instrument manufacturers recommend 12-month calibration intervals. The Condition Monitoring Manager can choose to send the instrument out of plant for third party calibration or set up an instrument test station for in-house calibration. Once the necessary test equipment and the instrument specific calibration procedure is in place, a complete check of the analyzer should only take about four hours.

A significant portion of the Analog circuitry common to most data collectors and analyzers has been covered up to this point, but all signal conditioning circuits, including bandpass filters, integration circuits, and other special functions should also be tested on a periodic basis. The examples provided in this appendix and the following Sample Acceptance Report can be used as a guide in developing an Operational Verification procedure for a specific data collector or FFT analyzer.

Sample Acceptance Test Report

Report Date_____ Model _____ S/N _____ Channel_____

Amplitude Accuracy (+/- 1.0% at 100 Hz)   Channel ____

Sens. Range	Input Level	Specification	Measured Level	Status
10.0 Volts	7.000 Vrms	6.930>7.070	_____	_____
5.0 Volts	3.500 Vrms	3.465>3.535	_____	_____
2.0 Volts	1.400 Vrms	1.386>1.414	_____	_____
1.0 Volts	0.700 Vrms	0.693>0.707	_____	_____

Amplitude Accuracy – Filter Flatness (+/- 1.0%)

A/R(Fmax)	.01	.15	.25	.35	.45	.55	.65	.75	.85	.99	Status
50 KHz	__	__	__	__	__	__	__	__	__	__	__
20 KHz	__	__	__	__	__	__	__	__	__	__	__
10 KHz	__	__	__	__	__	__	__	__	__	__	__
5 KHz	__	__	__	__	__	__	__	__	__	__	__

Frequency Accuracy (+/- 0.1%)

Input Frequency	Measured Frequency	Specification	Status
100.00 Hz	_____	99.90 > 100.10 Hz	_____

ICP Voltage, Current Test and Accelerometer Calibration

Volts	Milliamps	Measured Level	Units	Status
_____	_____	_____	_____	_____

Dynamic Range (=/>72 dB)

Sens Range	Max Level	Min Level	Difference(>72dB)	Status
10.0 Volts	_____	_____	_____	_____
5.0 Volts	_____	_____	_____	_____
2.0 Volts	_____	_____	_____	_____
1.0 Volts	_____	_____	_____	_____

Amplitude Linearity (+/- 1.0% or +/- 0.1 dB)

Sens Range	Input @ 100 Hz	Specification	Measured Level	Status
2.0 Volts	1.000 Vrms	0.990>.0100	_____	_____
2.0 Volts	0.1000 Vrms	3.465>3.535	_____	_____
2.0 Volts	0.01000 Vrms	1.386>1.414	_____	_____
2.0 Volts	0.001000 Vrms	0.693>0.707	_____	_____

Transfer Function, Gain & Phase Accuracy (+/- 0.5 dB and +/- 1.0 degree)

10 Hz A/R	100 Hz A/R	1.0 KHz A/R	10 KHz A/R	Status
___dB, ___deg	___dB, ___deg	___dB, ___deg	___dB, ___deg	_____

Test Equipment Used _____
Tested By _____ Next Test Due Date_____

APPENDIX I
Oil Analysis

Written by
Daniel P. Walsh
Laboratory Director
National Tribology Services, Inc.
5 Lakeland Park Drive, Peabody, MA 01960
Tel: (978) 531-6123 ext. 27   Fax: (978) 531-6522
Email: dan@natrib.com

As powerful as vibration analysis is, no one diagnostic or monitoring technique can be relied upon to assure continued operation of one's important machinery. For this reason, the author has elected to include this appendix on oil analysis to enhance the analyst's aresnal of available tools. Oil is the lifeblood of machinery, and analysis of it on a routine basis will similarly reveal a great deal of information about its health. This guide is aimed at providing a general introduction to the world of oil analysis, introduce the most common tests performed, and demonstrate how the technique is integral in the world of condition monitoring.

A Little History

Oil analysis as a tool for condition monitoring is not new. It has been used for at least fifty years in determining the wear condition of machinery. Railroad companies in the late 1940s and early 1950s found that the metals in a sample of used oil revealed the condition of the wearing parts in their locomotive engines. Likewise oil analysis has been used by the armed forces since the 1960s, and today they are the largest users in the world, monitoring the condition of everything from aircraft jet engines and helicopter gearboxes to armor vehicle powerplants. Outside of the military, construction equipment operators use the technique widely (Figure I.1), whereas industrial plants are slowly beginning to adopt the approach as an indispensable tool. Today, oil analysis is taking its place alongside vibration

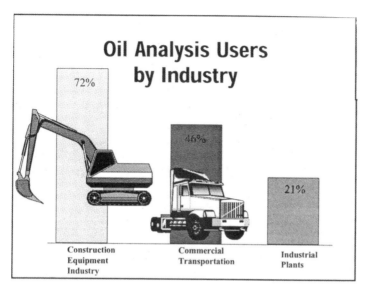

Figure I.1. Oil analysis users by industry. Industrial plants have been slow to adapt oil analysis because of the complexity and diversity of industrial equipment.

monitoring as an indispensable and valuable predictive maintenance tool in industry.

Oil analysis also plays an important role in the testing and start-up of new machinery. Early death syndrome is well known. Errors in assembly, improper tolerances, blocked lubrication ports, and factory contamination can all work to cause the early demise of the new machine. While monitoring brand new machines and machines which have recently "crashed" are well-recognized functions of oil analysis; it is the routine, periodic oil analysis which is the most valuable.

This guide aims to address the following details of the technology:

- examines the three facets of oil analysis,
- introduces the principal tests used
- guides setting up a program
- instructions on sampling
- discusses how alarms are developed and applied, with typical guidelines
- reviews management approaches to a successful program

The Three Facets of Oil Analysis

It is well worthwhile to review some basic tenets of oil analysis and its application to condition monitoring. There are many different types of tests that are used to evaluate lubricants. The tests specified must cover three areas: machine condition, contamination condition, and lubricant condition. (Figure I.2).

The purpose of which:

- To monitor for the presence of contamination detrimental to the oil and its application,
- To compare the chemicals and physicals of the oil against the virgin oil to determine if the oil is still an adequate lubricant, and
- To analyze for metal particles to determine the condition of the machine's oil wetted (wearing) parts.

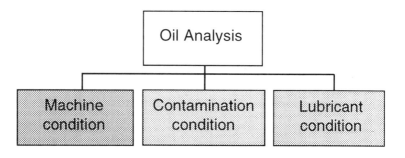

Figure I.3. Three components of oil analysis

Failure to test in all three areas may result in an unnoticed system failure, thus showing up a disregard for the equipment and the technology by both maintenance technicians and the laboratory. The tests are the most applicable to condition monitoring. Some may overlap the three areas of interest, but that is reassuring as it provides corroborating evidence of an abnormality.

Machine Wear Analysis

The wearing parts of a machine such as the gears, hydraulic pistons, bearings, and wear rings generate fine metal particles during normal operation. At the onset of a severe wear mode the particle size increases and the appearance of the particles change. Knowledge of particles and how they relate to the mode of wear (tribology) permits a trained analyst to determine the wear status in a machine by measuring the fine and coarse metal particles and then examining the particles under a microscope. The testing for wear metals for condition monitoring and predictive maintenance are tested predominantly in the following ways:

- Spectrochemical Analysis
- Wear Debris Analysis/Ferrography

Spectrometric Analysis: Technique for detecting and quantifying metallic elements in used oil resulting from wear, contamination, and additives. The oil sample is energized to make each element emit or absorb a quantifiable amount of energy, which indicates the element's concentration in the oil. The results reflect the concentration of all dissolved metals (from additive packages) and particulate (Figure I.3). This test is the backbone for all oil analysis laboratories today, as it provides information on machine, contamination and wear condition relatively quickly and accurately.

As important as the spectrometer is to oil analysis, it has a severe limitation. In its normal operation, it only sees dissolved metals and the finest debris, which are generally associated with benign or normal wear. Coarse particles are of greater interest because they usually are generated by severe wear modes. Particle detection efficiency is poor for particles 5 microns (μm) in size or greater (Figure I.4). Particles greater than 10 μm in major diameter are the result of abnormal, wear modes, and these particles must be quantified. The technique is accurate to 10%, although new equipment is now reporting within 3%.

Rotrode Filter Spectroscopy: First introduced in 1992, this spectrometric technique detects large or coarse wear metals and contaminants in a used oil sample. The atomic emission spectrometer used in this method suffers from the same particle limitation described above, but this weakness is overcome by the pre-filtering of the sample through one of the electrodes, called a rotrode. "Coarse" particles include all particulate up to 25 μm in size but exclude all additives. "Coarse" particles are especially important since these particles are the first indicators of abnormal wear situations. RFS provides a low cost efficient screen for ferrography, and is superior to DR (direct

Standard Elements Tested for By Wear Particle Analysis				
CONTAMINANT OR ADDITIVE	WEAR METALS	COOLANT OR ADDITIVE	OIL ADDITIVE	CONTAMINANT OR WEAR METAL
silicon	iron aluminum chromium copper lead tin nickle silver	sodium boron	zinc phosphorous calcium magnesium barium molybdenum	vanadium

Figure I.3. Standard elements tested by spectrometric analysis

read) ferrography because it detects ferrous, nonferrous, and contaminant elements (usually 12 elements). Detection efficiency of large material gets poorer as particle size increases above 25 μm diameter. Its accuracy range is within 15%.

Analytical/Diagnostic Tests

Analytical Ferrography: A technique, which separates magnetic wear particles from oil and deposits them on a glass slide known as a ferrogram. Microscopic examination permits characterization of

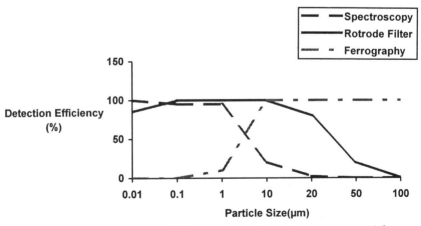

Figure I.4. Particle Detection efficiency comparison. Severe wear particles range from 10 μm upward.

Figure I.5. Ferrographic image of abnormal bearing wear from a large electric motor.

the wear mode and probable sources of wear in the machine. An automated version of this magnetic separation technique is DR (direct read) ferrography. It measures the ratios of large and small particles in the sample, and the data may be used to calculate the wear particle concentration and the severity index, two parameters, which allow for trending. It is an excellent indicator of abnormal ferrous wear occurring. It is unsuitable for nonferrous wear, however, and the test is most useful when a wear trend has been established. The ferrograph uses a magnetic separation technique which causes primarily ferrous particles to be deposited in a mono-layer on a glass substrate, segregated generally by size and into strings of particles. The resulting ferrogram is a permanent record of the ferrous wear in the machine.

Patch test A membrane filter patch is used to filter out debris from a known volume of oil. The membrane pore size is typically 0.45 μm, and a diluent is used to aid filtration. The test catches all debris, and is very effective if used in conjunction with ferrography. If spectroscopy shows that primarily non-ferrous metals are involved with a severe wear mode, the ferrogram effluent may be tested with a patch.

Both tests allow the particles of interest to be explored under a high powered microscope in order to identify the size, shape, metallurgy, and surface texture of the particles, all leading to an assessment of the wear condition in the machine and the source of the severe wear particles. These tests should only be done on an exception basis since they are time consuming and expensive. Fortunately, technology is available today with traditional spectroscopy plus RFS to quickly determine all the metals in the sample, both fine and coarse, so the decision whether to perform ferrography or patch test is easy.

Lubricant Condition Monitoring

Oil, being the lifeblood, needs to be in an acceptable condition in order to carry out its primary functions. Extending the drain interval of an oil (changeout time) is primarily based on the condition of the oil. To determine whether an oil sample is fit for further service, several tests may be specified, often comparing the used oil against a virgin oil of the same make and type.

Viscosity. The resistance of a fluid to flow. Viscosity is the most important lubricant physical property. Lubricants must have suitable flow characteristics to insure that an adequate supply reaches lubricated parts at different operating temperatures. The viscosities of lubricants vary depending on their classification or grade, as well as the degree of oxidation and contamination in service. If viscosity of the lubricant differs by more than 10% from nominal grade, the lubricant supplier recommends a change of oil. When the equipment is on a condition monitoring program, more specific controls may be put in place. Oil viscosity is expected to rise over time and use, and loss of viscosity is considered to be more serious than an increase. Therefore, a working alarm range is +20%, –10%, i.e., not more than 20% over nominal, and not less than 10% under nominal grade. Accuracy of this test is very good, at 0.5%. Viscosity is the most important physical property of a lubricant and should be checked periodically. An abnormal viscosity is an excellent indicator of a problem occurring, and should be acted upon.

Viscosity Index (VI) is an indication of how viscosity of the oil varies with temperature, and is called for when the machine must operate over a wide temperature range. Today's oils all have very good working temperature ranges, with synthetic oils having the highest viscosity indices.

Oxidation. Oil and its contents react with oxygen under a variety of conditions to form harmful by-products. These by-products are generally acidic in nature and are corrosive to the equipment and the oil itself. When the oil is oxidized, the oil must be replaced and disposed of. There are a number of tests used today to monitor for oxidation or acid. Total Acid Number (TAN), Total Base Number (TBN), Rotary Bomb, pH, and FT-IR are all being used routinely.

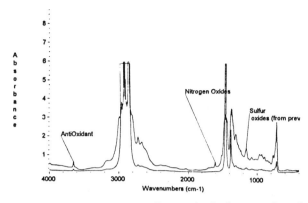

Figure I.6. FTIR compares "fingerprint" of new and used oils.

Infrared Analysis (FT-IR). Spectrometric technique for detecting organic contaminants, water and oil degradation products in a used oil sample. During a lubricant's service life, oxidation products accumulate, causing the oil to become degraded, and in most instances, slightly acidic. If oxidation becomes severe, the lubricant will corrode the equipment's critical surfaces. The greater the "oxidation number," the more oxidation is present. Similarly, the "nitration number" reflects the level of nitrogen compounds in the oil resulting from nitrogen fixation (common in natural gas fueled engines). Conditions such as varnishing, sludge deposits, sticky rings, lacquering, and filter plugging occur in systems with oxidation and/or nitration problems. Infrared spectroscopy also indicates contamination due to free water, glycol antifreeze, soot deposits, and fuel dilution. There are guidelines issued for oxidation numbers and liquid contaminants by manufacturers, but this is essentially a trending tool. Despite the advantages of FT-IR, certain ASTM tests should be routinely performed on certain critical machinery in order to conform to manufacturer's requirements, or to the user's standards.

Total Acid Number (TAN). A titration method designed to indicate the relative acidity in a lubricant. The acid number is used as a guide to follow the oxidative degeneration of an oil in service. Oil changes are often indicated when the TAN value reaches a predetermined level for a given lubricant and application. An abrupt rise in TAN would be indicative of abnormal operating conditions (e.g., overheating) that require investigation. Most lubricant suppliers give TAN condemnation limits in the bulletins. Accuracy of this test is within 15%.

Total Base Number (TBN). The converse of the TAN, this titration is used to determine the reserve alkalinity of a lubricant. The TBN is generally accepted as an indicator of the ability of the oil to neutralize harmful acidic by-products of engine combustion.

Contamination Control

Sources of Contamination

Solid contamination can be generated by many events so that high contamination levels at a given point in a machine can be used as a signal that something is abnormal.

- It could be caused by a failed filter, in which case two samples, taken upstream and downstream of the filter, can prove the failure. (That argues for installing sampling points on either side of the filter and monitoring just downstream on a regular basis. A high particle count will be followed by a sample from each side of the filter.)
- An increased wear rate in the wearing parts of the machine can increase the number (and size) of the particles present.
- Contamination can enter the system through a breather pipe, or through faulty seals, or because a protective filler cap was not replaced.
- When the oil becomes acidic or highly oxidized, it can attack the metallic components in the oil wetted path and generate a great deal of corrosive sediment.

These are but a few of the sources of solid contaminants. Spectroscopy will give a good indication of the elemental concentration of any debris present, provided that the material is not organic in nature. The most common form of contamination is moisture. This has many effects, including the aiding of acidity formation and lack of lubrication. The major tests for both particle contamination and moisture will be addressed here.

Particle Count (PC). A method used to count and classify particulate in a fluid according to accepted size ranges, usually to ISO 4406[1] and NAS 1638[2] There are several different types of instrumentation on the

---

[1] ISO 4406: International Standards Organization, 1994
[2] NAS 1638: National Aerospace Standard, 1956

market, utilizing a variety of measurement mechanisms, from optical laser counters to pore blockage monitors. Turbine and hydraulic systems ("clean systems") require particle counting, and should not exceed ISO 16/12 (NAS 8) in most cases, although some applications require more stringent limits. Whereas they are principally used to monitor the cleanliness levels in hydraulic or turbine systems, their application has now been extended to gearboxes and other plant equipment thanks to new methods of analysis. Significant increases in particle counts can indicate that a filter has failed, a contaminant is entering the system, or a higher wear rate has begun, to name a few. Particle Count depends on subsequent testing to diagnose the root cause of the higher counts.

Water. Usually not desirable in oil, water can be detected visually if gross contamination is present (cloudy appearance). Excessive water in a system destroys a lubricant's ability to separate opposing moving parts, allowing severe wear to occur with resulting high frictional heat. Water contamination should not exceed 0.25 % for most equipment, and not more than 100 ppm for turbine lube and control systems. There are several methods used for testing for moisture contamination, each with a different level of detection. They are summarized here.

Tests	Limit of Detection	Test Cost	Advantages	Disadvantages
Crackle	1000 ppm (0.1%)	low	A good field indicator, easy to perform.	Qualitative only, not suitable for trending.
FT-IR water	1000 ppm (0.1%)	low	Quantitative, good for trends, easy to perform.	Other liquid contaminants (glycol) present can confuse the spectrometer.
Centrifuge	1000 ppm (0.1%)	low	Still widely used for fuels testing and the BS&W tests.	Not as effective nowadays because of demulsifier additives.
Karl Fischer	10 ppm (0.001%)	high	The best test for low moisture levels. Very accurate. Required for turbine systems and transformers.	Unless lab has a nitrogen spurge, lubricant additives may cause interference. Need lab expertise to run correctly.

Figure I.7. Testing for moisture contamination.

Sampling

Insights Into Sampling

Taking the oil sample is critical, especially if the sample will be analyzed for coarse metal particles. These larger heavy particles tend to settle out quickly and are more readily removed in the filter. Therefore their distribution in a machine's oil system is not at all uniform. Taking the sample from the active oil stream, and in the same way each time, is required to get good trending results. The selection of the sampling point and method should be made with inputs from the lab. Each person who will take samples should go through a short course on taking samples. Most labs have specialists who will spend a day in your plant selecting sampling points and methods, training personnel, and recommending special valves and sampling techniques.

Sampling Techniques

There are generally three sampling methods used today. A vacuum pump with disposable plastic tube is the most common way to take a sample from a non-pressurized sump. An ordinary valve or special oil sample valve can be used to sample from a pipe or side of sump. Care must be taken with valves to flush out the dead leg which is always there. Also, one must be careful to avoid taking a sample from an area of no flow, such as at a sump wall or, worse, from the bottom of a sump where heavy debris settles and accumulates. A sample from the sump bottom will contain particles which are "ancient history" and are of no interest to the analyst. These old particles tell nothing about current wear; they can only alarm the analyst. A second good sample will be necessary to dispel the alarm and confusion. For some difficult sumps, a custom made "dip-stick" is used with the vacuum pump. The plastic tubing is attached to clips on the dipstick and the dipstick inserted into the sump. The dimensions of the dipstick, the "stand-off," and its shape will ensure that the sample will be taken each time from approximately the same point, and away from the walls and bottom.

Some special sample valves have a spring-loaded ball seat, which is unseated when a "needle" is inserted. The oil, if under pressure, flows through the needle and into the sample bottle. If no pressure is present, then a suction pump is used to "pull" the sample. These special valves come in different sizes and configurations, and can greatly increase the speed and convenience of taking a sample. They are particularly useful in dirty environments such as off-highway, highway diesels, mining, and dusty or dirty industrial sites.

Getting a Representative Sample Containing All Particle Sizes

Remember that the wear particles and contaminants are a separate phase in the oil and tend to settle out of the oil. Particles which are very small, on the order of 5 micrometers or less, stay suspended. These can be the only particles the lab sees if the person taking the sample does not capture a representative sample of the "working" oil in the system. Most labs do not test for metal particles larger than 5 micrometers anyway, unless they are using Rotrode Filter Spectroscopy. Some labs use the DR Ferrograph which measures coarse iron. Thus, it becomes especially important to capture a representative sample if the lab can measure the coarse metals. Without coarse metal analysis, many severe wear modes will be missed.

Using Your Senses, Postmortems, and Quality

Using Your Common "Senses"

Since a sample should be taken while the machine is running at some steady state, or shortly after shutdown, the resulting sample will be hot and any volatiles will be evident to the alert nose. Even after the sample is received in the lab, volatiles attack the nearest nose when the cap is removed. Fuel dilution, freon contamination, and other chemical contaminants can be clearly identified by the smell. Visually, you can also learn something, especially about the presence of water. Depending on the additives in the oil, water will sometimes quickly separate from the oil, so that you can see the water and oil separately if the sample bottle is clear plastic or glass. If the water forms a stable emulsion, the sample will appear milky or clouded.

Normally, wear particles and solid contaminants are not visible to the unaided eye—they are simply too small. However, when severe wear is underway the metal wear particles are sometimes large and plentiful enough to form a glittering deposit at the bottom of the bottle. Again, with a clear bottle, you will be able to see these particles. This event would be indicative of an ongoing severe failure mode.

With regard to feeling the oil by rubbing it between the fingers, the particles would have to be quite large (and therefore severe) to be felt. While using your common "senses" can be useful in detecting abnormal oil conditions, keep in mind that used lubricants are an identified carcinogen, and gloves should be used when handling used lubricants. Further, the smelling of oils should be avoided. Professional labs are strongly vented. Be careful!

Equipment Applications

Industrial equipment needs a combination of the above tests for condition monitoring. The following table is a summary of the test applicability.

Equipment	Spectro Analysis	Viscosity	FT-IR	Particle Count	Karl Fisher	Total Acid No.	Total Base No.	Rotrode Filter
Engines	R	R	R				R	A
Compressor	R	R	R		A	R		R
Gearboxes	R	R	R					R
Bearings	R	R	R	A				R
Hydraulics	R	R	R	R	A	A		R
Turbines	R	R	R	R	R	R		R
Motors	R	R	R					R

Code: R: Required test
    A: Advisable test, provides extra detail, particularly during problem solving

Figure I.8. Tests for Condition Monitoring

Alarming Methodologies

It is worth reviewing the two types of alarming methodologies that are employed in oil analysis.

Absolute alarms These are alarms based on manufacturers' recommendations and/or lubricant supplier technical bulletins. These alarms generally define working ranges or condemnation limits and are most applicable to lubricant and contamination condition. Extensive research is conducted to arrive at these limits, and they provide a good starting point for any analysis program. Absolute alarm limits matter greatly when warranties on new equipment are at issue. Failure to comply with the recommendations is often viewed as justification for not honoring such warranties.

Statistical Alarms Manufacturers' guidelines for alarm limits or general standards have the disadvantage that they are based on average operational and performance situations, which may not accurately reflect the actual conditions of a specific machine. This is particularly applicable to machine condition. Statistical alarms limits are based on gathering a small sampling of data from equipment, analyzing the distribution of that data, and using this statistical characterization to set specific alarm limits. Statistical trend analysis allows the identification of the equipment in greatest need of attention, thus allocating maintenance in an efficient way.

Combining Absolute and Statistical Alarms Effective oil analysis exception management relies on the combination of both types of alarms. The appendices to this paper give some test condemnation limits. It is worth noting the variation by manufacturer. It proves that different companies use different criteria for normal operation. The following illustration is an example of the alarm combination. The condemnation limit is the absolute alarm. Statistical trending, taking into account variability based on the sampling, contamination, make-up oil, etc., will develop the standard deviations. Departure from this normal variability signals genuine problems occurring. This is the earliest possible time to take action and head off problems. Failing this, as the trend approaches its warning limit, action such as changing or cleaning the oil, or inspection of the unit is required.

Establishing statistical alarms which provide the earliest possible warning without false alarms is a difficult task. Factors such as adding or changing oil, filter changes, and sampling technique can distort results. However, if the following foundation steps are employed, a very good oil analysis and lubrication management program will be the result.

Guidelines to Alarming Specific Equipment

Engines-Diesel and Automotive Engines are a great piece of equipment to start with since everyone is familiar with them and has to maintain at least one at some point in their lives. Engine design has

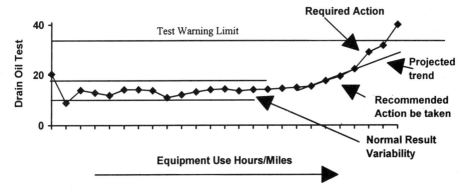

Figure I.9. The idealized graph shown is an example of how absolute and trendline alarms are used together. The test used could be iron content, viscosity, or other parameter. The normal result variability range takes into account minor variations caused by analytical accuracy, sample homogeneity, etc. As the trend approaches the warning limit, some action must be taken, either cleaning the oil or inspecting the equipment.

rapidly advanced in the last decade and that, combined with innovative uses of metals and ceramic components, has led to very low wear, and extended oil drain intervals. Engine manufacturers have recognized the advantages of oil analysis, and now produce bulletins on an annual basis, which summarize oil analysis alarms. They are also very helpful for engine component metallurgy knowledge. These bulletins may be had by calling your local engine distributor, or by contacting the oil analysis laboratory. Most engines use a multigrade oil, designed to provide protection over the wide thermal variations that this equipment sees. These lubricants contain the most advanced lubricant additive packages available, and both these packs and the viscosity are highly regulated by such authorities as the API and SAE. The following table summarizes condemnation limits for three popular engines.

	Engine Manufacturer		
Oil analysis test	Caterpillar (all models)	Cummins (all models)	Detroit Diesel (all models)
Spectroscopy, Iron	100 ppm	84 ppm	150 ppm
Spectroscopy, Copper	45 ppm	20 ppm	90 ppm
Spectroscopy, Lead	100 ppm	100 ppm	none specified
Spectroscopy, Aluminum	15 ppm	15 ppm	none specified
Spectroscopy, Chromium	15 ppm	15 ppm	none specified
Spectroscopy, Tin	20 ppm	20 ppm	none specified
Spectroscopy, Sodium	40 ppm	20 ppm	50 ppm
Spectroscopy, Boron	20 ppm	25 ppm	20 ppm
Spectroscopy, Silicon	10 ppm	15 ppm	none specified
Viscosity	+20 % to -10 % of nominal SAE grade	+/- 1 SAE grade or 4 Cst from new oil (Visc @ 100 degrees C)	+40 % to -15% of nominal grade (Visc @ 40 degrees C)
Water	0.25% max	0.2% max	0.3% max
TBN	1.0 mg KOH/g min value	2.0 mg KOH/g min., or one-half of new oil value, or equal to TAN	1.0 mg KOH/g min value
Fuel Dilution	5% max	5% max	2.5% max
Glycol Dilution	0.1% max	0.1% max	0.1% max
Ferrography	on exceptions	on exceptions	on exceptions

Notes: 1. These values represent all models of engine for the manufacturers listed and serve as a guide only. Consult your distributor for specific requirements.

Figure I.10. Condemnation Limits.

Compressors

Compressors come in many configurations, and are subject to certain machine condition and lubricant condition requirements. Reciprocating compressor lubrication often involves drip feeding high viscosity lubricant to the cylinders and use of a multigrade motor oil in the crankcase. Cylinder lubrication is once-through, and there is no easy way to determine proper lubrication unless the cylinders are visually examined. The crankcase oil is subject to many of the same considerations as diesel and gasoline engines described earlier.

Rotary compressor lubrication analysis is much more important, since the circulating oil experiences much wider temperature variations and also comes into contact with the gas being compressed. The oil is susceptible to oxidation at high temperatures, thereby increasing viscosity and acidity. High viscosity increases energy consumption, and the higher acidity promotes corrosion. The viscosity may also be subject to change based on the solubility of the gas being compressed.

Centrifugal compressors experience similar problems in the areas of acidity, viscosity, and wear metals due to extended prolonged service intervals. The following table gives a rough idea on the alarm limits for compressors. These values are subject to change.

Oil analysis test	Compressor Manufacturer	
	Carrier (Centrifugal, Rotary)	Atlas Copco (Centrifugal, Rotary)
Spectroscopy, Iron	15 ppm	none specified
Spectroscopy, Copper	500 ppm	none specified
Spectroscopy, Lead	15 ppm	none specified
Spectroscopy, Aluminum	15 ppm	none specified
Spectroscopy, Chromium	15 ppm	none specified
Spectroscopy, Tin	15 ppm	none specified
Spectroscopy, Zinc	500* ppm min	none specified
Spectroscopy, Nickel	15 ppm	none specified
Chlorine	20 ppm	none specified
Viscosity	+20 % to -10 % of nominal ISO grade	+20 % to -10 % of nominal ISO grade
Water	500 ppm max	0.2% max.
TAN	1.0 mg KOH/g max	

Notes: Many manufacturers specify their own lubricant, which carries its own condemnation limits.

 * This applies to the additive package in the lubricant and is a minimum ferrography should be performed on trend departures.

Figure I.11. Alarm limits for compressors.

Gearboxes

Most gear oil applications are concerned with the effects of shearing and high load bearing characteristics. Thus, wear metals and viscosity testing is of primary importance. Although all gear manufacturers give lubrication selection guidelines, most do not publish guidelines for wear condition. These applications rely heavily on baseline trends and statistical alarming techniques. Viscosity should increase over time and service because of oil and additive shearing, oxidation, and contamination. Some companies, such as Werner Pfleiderer, insist on keeping a close eye on viscosity variations. Others caution on effects of poor lubricant selection, and water contamination concerns.

Oil analysis test	Alarm limits	Further Action by Lab
Spectrochemical Analysis	10 % increase over last sample	
RFS (Doublecheck)	2:1 ratio or greater coarse to fine	Ferrography
Viscosity	+20%, -10% of nominal ISO grade	
Oxidation	0.2 Abs/0.1 mm over last sample	TAN
Water	0.25% max	Karl Fisher

Figure I.12.

Hydraulic Systems

Contamination is the most serious concern to hydraulic systems; they must be monitored periodically. Dirt and water are the most serious contaminants. It is estimated that 75 to 85% of all hydraulic system failures are a direct result of fluid contamination. Components such as pumps, valves, actuators, and conductors are affected by contamination in the following ways:

a) Increased internal leakage which lowers pump and motor efficiencies and reduces ability of valves to control flow and pressure accurately. This wastes horsepower and generates excess heat.

b) Corrosion of the system from acids that form due to fluid oxidation and water contamination.

c) Silt locking of valves due to particle contamination.

Guidelines for determining if a hydraulic fluid is unsuitable for service are listed here. Filtration and/or water removal is recommended when analysis alarms on particle count or water respectively.

Oil Analysis 315

Oil analysis test	Alarm limits	Further Action by Lab
Spectrochemical Silicon	15 ppm	
Spectrochemical Copper	12 ppm	
Spectrochemical Iron	26 ppm	
RFS (Doublecheck)	2:1 ratio or greater coarse to fine	Ferrography
Viscosity	+20%, -10% of nominal ISO grade	
Oxidation	0.4 Abs/0.1 mm over last sample	TAN (1.5 mg KOH/g max)
Particle Count*	17/14 ISO code	
Water	0.1% max	Karl Fisher

Note: Particle Count for many hydraulic systems is much lower.

Figure I.13.

Turbine and Circulating Oil

Turbine systems are full flow lubrication, requiring a large quantity of oil. The oils specified are usually mineral based, coming from highly refined basestocks. The components have little tolerance for contamination or oxidation, and these systems require frequent sampling. Manufacturers have laid out very good guidelines on how the oil must be maintained for trouble-free operation. Water contamination is particularly undesirable, as it tends to form an emulsion with the oil which in turn reduces the lubricity and induces corrosion. Heat transfer characteristics are also reduced which leads to elevated bearing temperatures. Therefore, water contamination control is usually very tight, and many filter systems contain coalescing filters. Particle contamination can clog lubrication ports, in-line filters and control systems. Turbine systems are required to maintain a low level of particle contamination. Viscosities of the oil tend to remain stable over many years, mainly because of the lubrication function. Oxidation is a concern and TAN condemnation limits are low compared to other types of equipment.

Oil analysis test	Steam turbines	Gas turbines
Spectroscopy, Iron	15 ppm	none specified
Spectroscopy, Copper	500 ppm	none specified
Spectroscopy, Lead	15 ppm	none specified
Spectroscopy, Aluminum	15 ppm	none specified
Spectroscopy, Chromium	15 ppm	none specified
Spectroscopy, Tin	15 ppm	none specified
Spectroscopy, Zinc	500* ppm min	none specified
Spectroscopy, Nickel	15 ppm	none specified
Chlorine	20 ppm	none specified
Viscosity	+20 % to -10 % of nominal ISO grade	+20 % to -10 % of nominal ISO grade
Water	500 ppm max	0.2% max
TAN	1.0 mg KOH/g max	

Figure I.14.

Steps to Good Program Management

1. Know Your Equipment

Many engineers and technicians do not know their equipment from a lubrication viewpoint. Information such as oil wetted component metallurgy, equipment loading conditions, and environment are very important, and contribute a great deal in solving problems shown up by oil analysis. Find the equipment maintenance manuals or call the project engineer who installed the machine. Keep the information close at hand in your file, and provide this information to the oil analysis laboratory.

2. Read Your Equipment Specifications

Many manufacturers of equipment have published specifications for lubricants and their maintenance. For example, all manufacturers of turbine systems and diesel engines have recommended guidelines with absolute limits for degradation, wear metals, and contamination. Many manufacturers of compressors and hydraulics have similar notes. Put these limits into your trending system and let the lab know. You now have absolute alarms and you stay within any warranties expressed by the equipment provider. The laboratory is happy because it has alarms specific to that piece of equipment.

3. Assess Your Lubricant

Oil analysis programs frequently find incorrect lube types and/or grades in equipment. A common problem recently is the use of new advanced synthetic lubricants, with different baseline properties than the original mineral based oil specified in the maintenance manual. Consult with your lubricant vendor and have a product application guide handy. Advise the manufacturer of the differences and ask your lubricant vendor to consult with them if there is a problem. Send the lab samples of all your new oil so that a baseline analysis is performed. Laboratories generally do not charge for this analysis. The values obtained are your baseline for trending lubricant properties.

4. Understand Oil Analysis Testing

Be familiar with the tests that have been very briefly described in this paper and of any other tests, as they are necessary. It is important to understand that all test methods are subject to some normal, predetermined variability, depending on the test. Viscosity testing has a normal variation of only 0.5%. In contrast, the total base number (TBN) test has a variation of 15%. A reported TBN value of 7.0 could actually represent a value of between 5.95 and 8.05. The accuracy ranges have been defined for the tests. Consult with the laboratory to account for variation in statistical trending analysis. The value of the analytical test as an effective alarm trigger decreases as the test variation increases.

5. Sample Carefully

Poor sampling can cause distortion of several hundred percent in sample data. Follow established guidelines for sampling, and sample from the same point each time. Air, oil, and filter changes and conditions can result in distorted alarms. Similarly, not recording or reporting topping up and/or bleed and feed operations have the effect of artificially changing wear, contamination, and additive levels without any real change in equipment or lubricant condition. These events must be recorded so as to take account of variations to trendlines.

6. Start Logging Operation Time

Statistical trending programs are not scientifically valid without including operation time. A standard trend alarm for wear metals and contaminants is to signal an alert when an increase of 10 ppm over a 10-hour period is found in any of the spectrochemical analysis readings. Engine users have traditionally tracked time, in order to determine drain intervals and schedule overhauls. Oil trend data for diesel engines is generally highly developed, with manufacturers specifying absolute limits and trend alarms. Rotating machinery users generally do not have tachometers fitted to equipment and, in cases of turbines, hydraulics, and large gearboxes, the oil is present since installation and the units are running continuously. Time-based sampling is the most common and effective method in these cases, especially when performed at regular intervals. Avoid random sampling, as their predictive value can be suspect. In all cases, record the date when sampling.

7. Establish Baseline Trends

Taking one sample from a piece of equipment that has had no previous history will not give you information about the wear trend of the machine and it is difficult to set alarms at this point. A monthly analysis for three consecutive months will establish a good wear trend. At this stage, a decision may be made on the sampling frequency of that equipment (with criticality factored in) based on the data received. This is an example of how the program manages both the equipment and itself by exception, thereby reducing the effort needed to maintain it by busy personnel.

Where Oil Analysis Fits into a Predictive Maintenance Program

Oil analysis generally does better than vibration monitoring in monitoring certain machines such as reciprocating machinery, slow rotating machinery, and hydraulic systems. But don't think that these applications limit oil analysis, or that they are even the most important applications. It is on fast rotating machinery where oil analysis may be most valuable, despite the fact that vibration monitoring also is well suited for these machines. The two technologies provide insights which complement and reinforce each other so that a clear maintenance decision is possible for these machine which are often the most critical in the plant.

Since oil analysis "sees" the machine from an entirely different perspective than vibration analysis, when the oil results confirm the vibration results, you can be more confident when taking specific corrective action suggested by the two analyses.

Production managers find it difficult to say "no" to a shutdown with such clear evidence as a color photograph of large fatigue chunks or cutting wear particles, especially when the vibration readings are revealing a high peak at gear mesh frequency.

Software

Recent software advances and an awareness of oil analysis in the vibration monitoring industry have combined to provide end users with more integrated software solutions for their vibration and oil analysis programs. Most vibration software packages offer modules for oil analysis data and most labs offer easy data transfer methods. The goal here is to provide the end user with the data and tools for him to manipulate the data for his own purposes. Windows-based software is now allowing for more exciting advances in this field, such as exceptional data trend analysis, and high resolution ferrographic images.

Overcoming Corporate Culture to Put Oil Analysis to Work

The biggest obstacle to oil analysis becoming an equal tool with vibration monitoring is entrenched corporate thinking regarding lubrication and its testing. Most maintenance personnel, even those very sophisticated in vibration monitoring, leave oil analysis out of their predictive maintenance thinking, or at best put it on the margin. The reason is that oil has usually been the domain of "the oiler" whose main goal was to ensure that sufficient amounts of the correct lubricants were maintained in each machine. The oil supplier, more often than not, also supplied "oil analysis" which gave the oiler and management the good feeling that the oil was O.K. However there are real problems in this.

- First, the testing is often not done on a regular periodic basis.

- Second, the results are typically returned so long after the sample is taken that its significance is lost.

- Third, the testing performed is mainly the lube chemicals and physicals, with insufficient attention paid to monitoring contamination and wear of the machine itself.

- Fourth, the results are not usually made available to the people who are doing the vibration monitoring.

- Finally, the company doing the testing is usually the lube manufacturer whose main objective is to sell lubricants, not to discover the machine is failing, the oil can continued to be used, or that the oil can be reclaimed by filtering and dewatering. When an independent lab is brought to the task, they can make recommendations unshackled by conflicting corporate interests.

The cost of oil analysis supplied by the large lubricant suppliers is generally little or nothing, at least on the surface. But what premium are you paying for the oil?

- Can you purchase equal lubes from the same or another supplier without an oil analysis program at a reduced price?
- What is the cost of a full dynamic program by an independent lab?

Changes to "business as usual" are required before oil analysis will occupy its rightful place in the field of predictive maintenance.

AFTERWORD
Not the End

Having reached the end of this book, it is time to point out to the reader that it is only the beginning of one's study of vibration analysis. A good understanding of the principles of modern FFT vibration analysis and monitoring is only the beginning of establishing competence in the field.

The reader must now begin to learn from actual experience and must gain an understanding of why a machine is designed and constructed the way it is. Why is the pump impeller cast rather than forged or designed as a weldment? Why is the base made of wide flange beams rather than channels or box beams? Is the bearing housing design adequate for the thrust loads of the application?

The vibration analyst must ask questions like these primarily because no one else will. One of the answers may well be the key to solving the problem. Listen to the answers.

With regard to monitoring machinery, remember that almost any fool can tell the client or employer that a machine has failed again. Many can tell him that the machine is about to fail in time to avert disaster. Only a few are capable of finding out why the machine fails so often. Try to join the ranks of these competent few.

Having trained hundreds of technicians to do vibration analysis, the author has found that experience is gained quite quickly in the machinery vibration field if the problems are always attacked in the same logical way:

1. Observe the design and operation of the machine. Touch things to see what vibrates and what is hot.

2. Learn about the operating history of the machine. Have operating conditions changed recently? Has the machine been rebuilt or the process changed in any way? Does it fail often?

3. Ask people stationed around the machine for their opinion. If a janitor takes his coffee break near a compressor, he can tell you that the machine has been making more noise recently. He will also lack any prejudice as to the cause of the problem.

4. Make a hypothesis as to the cause of the problem (but keep it a secret to avoid being embarrassed when you are wrong).

5. Run tests to prove or disprove the hypothesis.

6. Use this information to modify your hypothesis.

7. Repeat the hypothesis/test loop until the hypothesis has been proven. Often, only about three iterations of the loop are required for success.

8. If the problem has several possible causes that cannot be separated out by testing, recommend the simplest fixes first. There is always time to "throw money" at a problem after the easy fixes have been tried.

If the above course of action has been diligently tried but success has eluded you, realize that you have probably done as well as anyone else would have. It may then have become time to bring in an outside consultant—not because of his brilliance, but because of his different perspective. There is no dishonor in this.

Good Luck!

Additional Questions

The following fifteen questions are like a well given take-home test. No amount of additional time will help you determine the right answers. They have been designed to confuse and befuddle you. It will take a thorough knowledge of vibration analysis (which you have surely gained by now) to properly answer them. Good luck.

1. Each of the following techniques requires motion data (such as from an accelerometer) as an input signal to a spectrum analyzer plus another input signal or piece of equipment beyond the accelerometer and spectrum analyzer. List which items go with each technique:

TECHNIQUE	REQUIRED ITEMS
A. Transfer Function	1. Low pass filter between analyzer and acceleromter
B. Synchronous Time Averaging	2. 6 DB/OCTAVE or 12 db/octave low pass filter
C. Order Tracking	3. Two phase matched microphones (no motion signal)
D. Coherence Determination	4. Microphone and single integration of accelerometer data
E. Acoustic Intensity	5. Second channel of data (often Force data)
F. Analog Integration	6. Tack pulse (once/rev)
G. Out-of-band Signal removal	7. Pulse multiplier and Tack pulse (once/rev)
	8. Low pass tracking filter

2. To view the effect of one particular gear of a gearbox, what signal processing technique is required?
   a. Zoom (Frequency Expansion)
   b. Synchronous time averaging
   c. Order tracking
   d. Acoustic intensity
   e. none of the above

Answers begin on page 325

3. To determine which of several gearboxes in close proximity to each other on a machine room floor is the major noise source, which technique may be used?
   a. Compare the airborne noise spectra to the known forcing frequencies of each gearbox.
   b. Synchronous time average the airborne noise signal, triggering off successive gearbox shafts.
   c. Measure coherent output power between airborne noise and the vibration of successive gearboxes.
   d. All of the above
   e. none of the above

4. When order tracking, can digital integration of an accelerometer signal be used to obtain velocity data?

5. Power Spectral Density is used to compensate for changes in filter bandwidth (such as results from different analyzer frequency settings) when examining:
   a. Pure tones
   b. Broad band noise
   c. Frequency varying signals
   d. All of the above

6. What is the roll-off rate of a single integration (i.e. conversion of acceleration to velocity or velocity to displacement)?

7. What is the roll-off rate of a double integration (i.e. conversion of acceleration to displacement)?

8. Which of the following *cannot* be used to balance a kraft paper machine dryer running in the machine at a speed of 60 RPM (1 Hz)?
   a. Accelerometer
   b. Velocity pickup
   c. Proximeter
   d. Velometer

9. What signal processing technique requires a tracking low pass filter for antialiasing protection, and a way to bypass the spectrum analyzer's antialiasing filter?
   a. Transfer function (dual channel analysis)
   b. Synchronous time averaging
   c. Order tracking
   d. Zoom (Frequency Translation)
   e. Acoustic intensity

10. If the vibration of a speed varying machine is averaged in a spectrum analyzer without benefit of order tracking, which of the following is likely to occur?
    a. Improper amplitude
    b. Peaks at unidentifiable frequencies
    c. Frequency smearing
    d. All of the above

11. Which phase angle *cannot* be used for balancing?
    a. The phase readout obtained by synchronous time averaging, triggering off the shaft of interest
    b. The phase readout obtained from a strobe-light type balancing analyzer
    c. Transfer function phase, where one of the vibration signals is from the shaft of interest

12. Ceptstrum is most closely related to:
    a. Power spectrum
    b. Coherent output power
    c. Transfer function
    d. Cross correlation
    e. Auto correlation

13. The inability to see an expected peak in a summation averaged spectra could *not* have resulted from:
    a. The peak doesn't exist
    b. The peak is buried in broad band noise
    c. Insufficient dynamic range of the spectrum analyzer or tape recorder
    d. Broken transducer or cable
    e. Insufficient averaging
    f. Varying phase relationship to accelerometer location

14. Cavitation can appear in an acceleration spectra as:
    a. Low frequency noise
    b. Broad band noise
    c. Pure tones at discrete frequencies
    d. All of the above
    e. None of the above

15. A clear indicator of the reliability of an impact test for the determination of natural frequency is:
   a. Transfer function
   b. Coherence
   c. Power spectral density
   d. Coherent output power
   e. Cross correlation

ANSWERS

Question 1

Answers	Reference
A - 5	Figure 5.8, page 111 (Chapter 5)
B - 6	Page 40 (Chapter 2)
C - 7 and 8	Page 36–39 (Chapter 2)
D - 5	Page 117–118 (Chapter 5)
E - 3 or 4	Page 201–203 (Chapter 9)
F - 2	Page 60 (Chapter 3)
G - 1 or 2	Page 60–61 (Chapter 3)

Question 2

Answer: b. using the shaft of interest for triggering
   a. won't work because every gear mesh frequency involves two gears, not one
   c. irrelevant for a constant speed problem
   d. won't yield adequate spacial resolution because every gear mesh frequency involves two gears, not one

Reference: page 40 (Chapter 2)

Question 3

Answer: d.

Reference: For answer a., see page 87–90 (Chapter 4). Note that this method requires intimate knowledge of each gearbox, and assumes that each gearbox has different mesh frequencies. For answer b., see page 40 (Chapter 2). For answer c., see page 121 (Chapter 5)

Question 4

Answer: No, since digital integration requires mathematical division by a frequency ramp (1/omega), and frequency is not defined when order tracking. The number of orders is defined, a forcing frequency value would vary from average to average as shaft speed changes.

Reference: page 36–39 (Chapter 2)

Question 5

Answer: b. because this is the only signal of the three that will vary in amplitude as the bandwidth of the filter changes.

Reference: page 176–178 (Chapter 8)

Question 6

Answer: 6 db/octave

Reference: page 60 (Chapter 3)

Question 7

Answer: 12 db/octave

Reference: page 60 (Chapter 3)

Question 8

Answer: b. because a velocity pickup rolls off below about 10 Hz, and is therefore blind to a signal at 1 Hz. An accelerometer can be used, but may require a low pass filter if there is so much gear mesh energy that there is not enough spectrum analyzer dynamic range to see the low frequency, low energy unbalance in the presence of the high frequency, high energy gear mesh. A velometer is an accelerometer with a 6 db/octave roll-off low pass filter, and a proximeter reads down to 0 Hz.

Reference: page 51–57 (Chapter 3)

Question 9

Answer: c.

Reference: page 36–39 (Chapter 2)

Additional Questions 327

Question 10

Answer: d.

Reference: page 36–39 (Chapter 2)

Question 11

Answer: c. Transfer function phase cannot be used for balancing because it shows the phase relation between two dynamic signals, not between the signal of interest and a known, fixed reference point.

Reference: page 40–41 (Chapter 2) and page 179–180 (Chapter 8)

Question 12

Answer: e.

Reference: page 203–205 (Chapter 9)

Question 13

Answer: f. In summation averaging, the phase relation of a particular peak with respect to accelerometer location is irrelevant. All of the other listed causes are possible.

Reference: page 34–35 (Chapter 2)

Question 14

Answer: d. Both a. and b. are true by the nature of the bubble implosion, c. is true if the wide band noise excites a specific natural frequency of the machine being measured.

Reference: page 94 (Chapter 4)

Question 15

Answer: b.

Reference: page 119 (Chapter 5)

Index

A

A/D converter, 29
acceleration, 6–8, 55–57
accelerometers, 55–57
acceptability criteria, 21–23
acoustic intensity, 201–203
air gap, 285
all pass filters, 11–12
amplification ratio, 113
amplitude, 1–4
amplitude accuracy, 290–291
amplitude linearity, 67
analog tape recorders, 166
analysis range, 34
antialiasing, 28–29, 38–39, 157
autocorrelation, 192–195
autorange, 157
average cross spectrum, 178–179

B

background readings, 79–80
balancing, 243–254
band pass filters, 13
band-limited white noise, 117
base design, 82
base problems, 81–82
baseline criteria, 21, 139–141
bearing faults, 90
bearings, 90–93
blade frequencies, 90
buffer memory, 31

C

calibration, 67–70
cavitation, 94, 216–218
cepstrum, 203–205
chirp, 117

coherence, 113, 117–121, 184–186
coherent output power, 121–122, 186
CPM, 3
complex numbers, 169–172
constant percent filters, 15–17
constant width filters, 17
continuous monitoring, 130–131, 148
corner frequency, 14
critical damping, 105, 173
cross channel properties, 108–122
cross correlation, 196–199
cumulative distribution function, 199–201

D

damping, 102–106, 113
damping ratio, 102–106, 113
decibels, 9–11
diagnostics, 134
digital oscilloscope, 236
digital tape recorders, 168
digitization, 29–30
displacement, 2–3, 49–51
dual channel spectrum analysis, 42–43, 108–122
DuHammel's intergral, 188
dynamic range, 30–31, 46–47, 60–61, 157, 294

E

equations of motion, 104–110
exponential averaging, 35

F

Faltung intergral, 188
ferrography, 302–303

330 Index

FFT devices, hand held, 155–160
FFT spectrum analyzer, 34, 154–155, 234
filter flatness, 292
filters for diagnosis, 17–18
filters, characteristics, 11–17
flexible couplings, 85–86, 227
flexible rotors, 253–254
fluid film bearings, 93
force hammer, 63–64, 112, 213
force transducers, 63
force window, 183
forcing frequencies, 83
frequency, 1–4
frequency accuracy, 291
frequency analysis, 20–21
frequency expansion, 41–42, 88–89
frequency range, 34, 66–67, 157–158

G

gear tooth error, 87–88
generators, 269–286
gear/sensor pulse demodulation, 230
gears, 87–90, 227–228
ghost frequencies, 89
ground loops, 82

H

Hertz, 3
high pass filters, 13
Hoodwin torsional accelerometer, 231
hunting tooth frequency, 89
hydrophones, 65–66

I

impulse, 211
impulse response, 186–189
impulsive testing, 112, 114, 212–216
influence coefficients, 252
infrared analysis, 305
input attenuation, 27–28
input time, 172–174
instantaneous spectrum, 174–175
integration, 57–60
inverse transfer function, 189–191

J

Jericho, the walls of, 220–222

L

log decrement, 105–106, 173–174
logarithms, 9
looseness, 94–95
low pass filters, 12
low speed rotation, 228–229

M

mass-spring system, 6
mesh frequency, 87–89
microphones, 63–65
misalignment, 84–86
MMF, 285
modal analysis, 122–124
monitoring interval, 138
monitoring matrix, 144–150
monitoring non-vibration parameters, 138–139
monitoring program, machine selection, 135–138
motors, 95, 269–286
multiplane balancing, 252–253
multi-degree of freedom system, 110–112

N

natural frequencies, 99–100, 107–108
Nyquist plot, 182

O

octave and 1/3 octave meters, 152–153
octave filter, 15–17
one-third octave filter, 15–17
optical transducers, 230–231
order tracking, 36–39
overalapped processing, 45–46
oxidation, 304

P

P-V diagrams, 235–236
particle count, 306
pass band, 13–14
patch test, 303
peak, 3
peak hold (averaging), 35
peak-to-peak, 3
period, 3, 5–6
periodic condition monitoring, 131–133, 148–149

Index

phase, 1, 4–5, 8–9
pink noise, 117
piping problems, 80–81
piston-crank mechanism, 2–3, 7–8
polar plot, 182
power spectral density, 176–178
power spectrum, 175–176
predictive maintenance, 128
pressure transducers, 65–66
probability density function, 199–201
proximeters, 51–54
PSD, 176–178
pulse theory, 211–222

Q

Q, 105–106

R

range translation, 41–42, 89
real time bandwidth, 44–45
reciprocating equipment, 223, 225–226, 233–242
recording devices, 165–168
reject band, 12–14
resonance, 99–100
response window, 184
rigid rotors, 253
RMS, 3–4
RMS spectrum, 175
rolling element bearing, 90–92, 194, 218–220
rotational speed, 84–85
rotrode filter spectroscopy, 301
rubs, 94–95

S

sampling, 36–41
sensitivity constant, 66
severity chart, 22
shaft bending, 84–85
shaft position encoders, 231
shock pulse method, 218–220
sidebands, 86, 88, 226
signal-to-noise ratio, 34
sine sweep, 117
sine wave, 1–4
single plane balancing, 249–252
single-degree-of-freedom, 104–106, 224–225

slip frequency, 95
slot frequency, 276–279, 282–285
sound, 23–24
spectrometric analysis, 301
spike energy method, 218–220
strain gauges, 229–230
summation averaging, 34–35
synchronous time averaging, 40–41, 236

T

temperature linearity, 67
time averaging, 35
time window, 32–34, 43–44
torsional vibration, 223–232
total acid number, 305
total base number, 306
transducers location, 78–79, 141–142
transducer mounting, 61–63
transducer selection, 142–143
transducer specifications, 66–67
transducers, 49–77
transfer function, 108–117, 179–182
transmissbility, 179
trending, 133
trending software, 160–163
trouble shooting guide, 79–83
tunable filter meters, 153–154
two channel analyzer, 42–43

U

unbalance, 84
untrendable failures, 133–134

V

vectors, 248–249
velocity, 6–7, 55–57
velocity pickups, 54–55
vibration defined, 1–3
viscosity index, 304

W

whirl, 93
white noise, 116–117
weighting, 32–33, 177–178

Z

zoom, 41–42, 89